Advanced Applications for Smart Energy Systems Considering Grid-Interactive Demand Response

Advanced Applications for Smart Energy Systems Considering Grid-Interactive Demand Response

Special Issue Editor

Angèle Reinders

MDPI • Basel • Beijing • Wuhan • Barcelona • Belgrade

MDPI

Special Issue Editor
Angèle Reinders
University of Twente and Eindhoven University of Technology
The Netherlands

Editorial Office
MDPI
St. Alban-Anlage 66
4052 Basel, Switzerland

This is a reprint of articles from the Special Issue published online in the open access journal *Applied Sciences* (ISSN 2076-3417) from 2018 to 2019 (available at: https://www.mdpi.com/journal/applsci/special_issues/Smart_Energy_Systems).

For citation purposes, cite each article independently as indicated on the article page online and as indicated below:

LastName, A.A.; LastName, B.B.; LastName, C.C. Article Title. *Journal Name* **Year**, *Article Number*, Page Range.

ISBN 978-3-03921-998-8 (Pbk)
ISBN 978-3-03921-999-5 (PDF)

Cover image courtesy of Angèle Reinders.

Contents

About the Special Issue Editor

Angèle Reinders received an MSc in Experimental Physics at Utrecht University (1993), where she also received her Ph.D. degree (1999) in chemistry. At present she is a full professor in Design of Sustainable Energy Systems at Eindhoven University of Technology and an associate professor at University of Twente in the Netherlands. In the past she conducted research at Fraunhofer Institute of Solar Energy in Freiburg, the World Bank in Washington D.C., ENEA in Naples, Center of Urban Energy in Toronto, and in various parts of Indonesia, and she was a Professor of Energy-Efficient Design at TU Delft. Based on these experiences, she developed a new approach towards energy research that is design-driven in scope. Since 2017, she has been the chair of the EU COST Action PEARL PV on the performance of PV systems, and she is conducting projects on smart energy systems, including the CESEPS project on Co-Evolution of Smart Energy Products and Services (2016–2019). She is known for her books "The Power of Design—Product Innovation in Sustainable Energy Technologies" (2012) and "Photovoltaic Solar Energy From Fundamentals to Applications" (2017) and for her involvement in the international IEEE PVSC conference, which she chaired in 2014 and 2017. In 2010, she co-founded the IEEE Journal of Photovoltaics, for which she serves as an Editor. She is also involved in various tasks of the Cover Photography International Energy Agency PVPS program, including Task 1 on Strategic PV Analysis and Outreach and Task 17 on PV for Transport.

applied
sciences

MDPI

Editorial

Special Issue on Advanced Applications for Smart Energy Systems Considering Grid-Interactive Demand Response

A.H.M.E. (Angèle) Reinders [1,2]

[1]   Department of Design, Production and Management, Faculty of Engineering Technology, University of Twente (UT), PO Box 217, 7500 AE Enschede, The Netherlands; a.h.m.e.reinders@utwente.nl
[2]   Energy Technology Group, Department of Mechanical Engineering, Eindhoven University of Technology (TU/e), PO Box 513, 5600 MB Eindhoven, The Netherlands

Received: 19 September 2019; Accepted: 19 September 2019; Published: 30 September 2019

1. Introduction

The continuous growth of our fossil fuel demand for energy consumption and the related increase in CO_2 emissions [1,2] are essential motivations to conduct research on smart energy systems which are able to reduce the emission of greenhouse gases. Namely, to strengthen the global response to the threat of climate change, as stated by the Paris Agreement of the United Nations [3], the global temperature increase should stay below 2 °C, or, preferably below 1.5 °C. Because fossil fuel-related CO_2 emissions for energy purposes have increased consistently over the last 40 years, reaching 69% of global GHG emissions in 2010, an urgent transition to more sustainable, smart energy systems is required to stay below this temperature threshold of 1.5 °C. This sustainable energy transition will involve a huge effort by a varied group of stakeholders, including policymakers, industry, business developers, the public, engineers and also scientists. In this context, societal acceptance, technological development, and suitable market design are vital to realizing this goal. In particular, this applies to sustainable, smart energy systems at a human scale, which can be found in households and in neighborhoods, where people live and interact with their energy systems. These sustainable energy systems are called smart grid households or smart grid pilots if they are integrated into a neighborhood.

Each of the nine chapters in this book relates to this framework and was previously published in a Special Issue of the journal *Applied Sciences* on 'Advanced Applications for Smart Energy Systems Considering Grid-Interactive Demand Response'. As such the chapters in this book report about interdisciplinary research results that combine technical, social, environmental, and economic aspects of sustainable, smart energy systems. For this Special Issue, authors were invited to submit manuscripts covering applied research on smart energy systems, smart grids, smart energy homes, smart energy products and services, and advanced applications thereof, in the context of demand response and grid interactions. Therefore, in this book, interesting results are presented based on the evaluation of real-life cases, energy pilots, prototypes of smart energy products, as well as end-user surveys and interviews.

2. Three-Layer Model for Assessments of Residential Smart Grids

This chapter presents an important framework, called the three-layer model, for the evaluation of a smart grid environment [4]. This three-layer model comprises three specific categories, or 'layers', namely, the stakeholder, market, and technology layers. Each layer is defined and explored, using an extensive literature study regarding their key elements, their descriptions, and an overview of the findings from the literature. The assumption behind this study is that a solid understanding of each of the three layers and their interrelations will help in a more effective assessment of residential smart grid pilots in order to better design products and services and deploy smart grid technologies in networks. Based on this review, it is concluded that, in many studies, social factors associated with smart grid pilots, such as markets, social acceptance, and end-user and stakeholder demands, are most commonly defined as uncertainties and are, therefore, considered separate from the technical aspects of smart grids. As such, it is recommended that, in future assessments, the stakeholder and market layers should be combined with the technology layer so as to enhance interaction between these three layers, and to be able to better evaluate residential smart energy systems in a multidisciplinary context.

3. Empowered Users

The active involvement of users in smart grids is often seen as key to the beneficial development of smart grids. In her chapter, Van Mierlo [5] investigates the diverse assumptions about how and why users should be active and to what extent these assumptions are supported by experiences in practice. This chapter presents the findings of a systematic literature review on four distinctive forms of user involvement in actual smart grid projects: demand shifting, energy-saving, co-design, and co-provision. The state-of-the-art knowledge reflects a preoccupation with demand shifting in the actual smart grid development. Little is known about the other user roles. More diversity in the types of projects regarding user roles would improve the knowledge base for important decisions defining the future of smart grids.

4. Indicators for Good System Design

Another chapter [6] frames the impact of various smart grid technologies, to provide a transparent set of indicators for the performance assessment of residential smart grid demonstration projects. The analysis comprises an evaluation of measured energy data of 217 households from three smart grid pilot projects in the Netherlands and a public dataset with smart meter data from 70 households as a reference. The datasets were evaluated for one year and compared, to provide insights into technologies and other differences based on seven key performance indicators, giving a comprehensive overview of system performance: monthly electricity consumption (100–600 kWh) and production (4–200 kWh); annually imported (3.1–4.5 MWh) and exported (0.2–1 MWh) electricity; residual load; peak of imported (4.8–6.8 kW) and exported (0.3–2.2 kW) electricity; import simultaneity (20–70.5%); feed-in simultaneity (75–89%); self-sufficiency (18–20%); and self-consumption (50–70%). It was found that the electrification of heating systems in buildings using heat pumps leads to an increase in annual electricity consumption and peak loads of approximately 30% compared to the average Dutch households without heat pumps. Moreover, these peaks have a high degree of simultaneity. To increase both the self-sufficiency and self-consumption of households, further investigations will be required to optimize smart grid systems.

5. Conclusions

In this book, the reader will further find interesting information about the design of smart energy products [7], about the transaction mechanism for energy pricing for prosumers [8], about ancillary services from residential heat pumps on the reserve market [9] and more. As such, we can conclude that this book provides an interesting overview of concurrent research on innovations into sustainable, smart energy systems, which is essential and indispensable for lowering the energy-related CO_2 emissions of our society. We wish our readers a good, inspirational reading experience.

Conflicts of Interest: The author declares no conflict of interest.

References

1. IEA (International Energy Agency). International Energy Agency, Statistics. Available online: https://www.iea.org/statistics (accessed on 19 August 2019).
2. IPCC (Intergovernmental Panel on Climate Change). Global Warming of 1.5 °C. An IPCC Special Report on the Impacts of Global Warming of 1.5 °C above Pre-industrial Levels and Related Global Greenhouse Gas Emission

Pathways, in the Context of Strengthening the Global Response to the Threat of Climate Change, Sustainable Development, and Efforts to Eradicate Poverty. Available online: https://www.ipcc.ch/sr15/chapter/spm/ (accessed on 30 September 2019).

3. UN (United Nations). *Adoption of the Paris Agreement, Document FCCC/CP/2015/L.9/Rev.1*; UNFCCC: Bonn, Germany, 2015; p. 32.

4. Reinders, A.; Übermasser, S.; Van Sark, W.; Gercek, C.; Schram, W.; Obinna, U.; Lehfuss, F.; Van Mierlo, B.; Robledo, C.; Van Wijk, A. An exploration of the three-layer model including stakeholders, markets and technologies for assessments of residential smart grids. *Appl. Sci.* **2018**, *8*, 2363. [CrossRef]

5. Van Mierlo, B. Users Empowered in Smart Grid Development? Assumptions and Up-To-Date Knowledge. *Appl. Sci.* **2019**, *9*, 815. [CrossRef]

6. Gercek, C.; Schram, W.; Lampropoulos, I.; Van Sark, W.; Reinders, A. A Comparison of households' energy balance in residential smart grid pilots in the Netherlands. *Appl. Sci.* **2019**, *9*, 2993. [CrossRef]

7. Sierra, A.; Gercek, C.; Übermasser, S.; Reinders, S. Simulation-supported testing of smart energy product prototypes. *Appl. Sci.* **2019**, *9*, 2030. [CrossRef]

8. Oh, E.; Son, S.-Y. Transaction Mechanism Based on Two-Dimensional Energy and Reliability Pricing for Energy Prosumers. *Appl. Sci.* **2019**, *9*, 1343. [CrossRef]

9. Posma, J.; Lampropoulos, I.; Schram, W.; van Sark, W. Provision of Ancillary Services from an Aggregated Portfolio of Residential Heat Pumps on the Dutch Frequency Containment Reserve Market. *Appl. Sci.* **2019**, *9*, 590. [CrossRef]

Review

An Exploration of the Three-Layer Model Including Stakeholders, Markets and Technologies for Assessments of Residential Smart Grids

Angèle Reinders [1,2], Stefan Übermasser [3], Wilfried van Sark [4], Cihan Gercek [1,*], Wouter Schram [4], Uchechi Obinna [5], Felix Lehfuss [3], Barbara van Mierlo [6], Carla Robledo [7] and Ad van Wijk [7]

[1] Department of Design, Production and Management, Faculty of Engineering Technology, University of Twente, P.O. Box 217, 7500 AE Enschede, The Netherlands; a.h.m.e.reinders@utwente.nl or a.h.m.e.reinders@tue.nl
[2] Energy Technology Group at Mechanical Engineering, Eindhoven University of Technology, P.O. Box 513, 5600 MB Eindhoven, The Netherlands
[3] Austrian Institute of Technology, Giefinggasse 4, 1210 Vienna, Austria; stefan.uebermasser@ait.ac.at (S.Ü.); felix.lehfuss@ait.ac.at (F.L.)
[4] Copernicus Institute of Sustainable Development, Utrecht University, P.O. Box 80.115, 3508TC Utrecht, The Netherlands; W.G.J.H.M.vanSark@uu.nl (W.v.S.); w.l.schram@uu.nl (W.S.)
[5] Van Hall Larenstein University of Applied Sciences, PO Box 1528, 8901 BV Leeuwarden, The Netherlands; uche.obinna@hvhl.nl
[6] Knowledge, Technology and Innovation, Wageningen University and Research, P.O. Box 8130, 6700 EW Wageningen, The Netherlands; barbara.vanmierlo@wur.nl
[7] Energy Technology Section, Department of Process and Energy, Delft University of Technology, Leeghwaterstraat 39, 2628 CB Delft, The Netherlands; c.b.robledo@tudelft.nl (C.R.); avanwijk@xs4all.nl (A.v.W.)
*   Correspondence: c.gercek@utwente.nl; Tel.: +31-534897875

Received: 23 October 2018; Accepted: 20 November 2018; Published: 23 November 2018

Abstract: In this paper, a framework is presented for the evaluation of smart grid environment which is called the three-layer model. This three-layer model comprises three specific categories, or 'layers', namely, the stakeholder, market and technologies layers. Each layer is defined and explored herein, using an extensive literature study regarding their key elements, their descriptions and an overview of the findings from the literature. The assumption behind this study is that a solid understanding of each of the three layers and their interrelations will help in more effective assessment of residential smart grid pilots in order to better design products and services and deploy smart grid technologies in networks. Based on our review, we conclude that, in many studies, social factors associated with smart grid pilots, such as markets, social acceptance, and end-user and stakeholder demands, are most commonly defined as uncertainties and are therefore considered separately from the technical aspects of smart grids. As such, it is recommended that, in future assessments, the stakeholder and market layers should be combined with the technologies layer so as to enhance interaction between these three layers, and to be able to better evaluate residential smart energy systems in a multidisciplinary context.

Keywords: smart grids; electricity market; flexibility; stakeholders; end-users; renewable energy; energy products and services

1. Introduction

For the successful deployment of residential smart grids, it is evident that interdisciplinary information about energy technologies, energy markets and the needs of various types of stakeholders must be identified, merged, and implemented in practice. Therefore, in this paper, the results of an in-depth literature study are presented, aimed at elaborating a framework for gathering knowledge and developing understanding about residential smart energy systems. This framework, called the three-layer model, comprises three layers: stakeholders, the market, and technologies [1]. Each layer is defined below. The framework originates from the European research program, ERA-Net Smart Grids Plus, in which it is considered to provide a common context for interdisciplinary smart grid research (see Figure 1) [2]. According to the International Energy Agency (IEA), a smart grid is defined as "an electricity network that uses digital and other advanced technologies to monitor and manage the transport of electricity from all generation sources in that (local) network to meet the varying electricity demands of end-users" [3].

In this paper, we focused on the specific category of smart grid environment known as residential smart grids and their pilot projects for experimenting with diverse features. Residential smart grids are located in the low voltage grid, usually in the built environment, and involve tens to hundreds of households that are equipped with smart energy products and services.

Our definition of smart energy products and services (SEPS) includes all the products and services that have the ability to support the active participation of end users by efficiently and reliably managing their energy systems and balancing the mismatch between electricity demand and supply [4].

1.1. Layers

Below each layer of the model is briefly described in order to provide a general understanding for the reader.

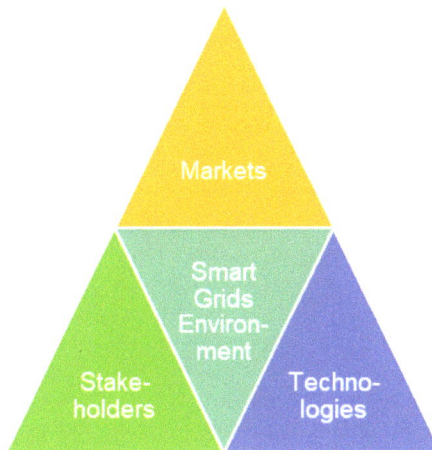

Figure 1. Three-layer research model for smart grids environments (adapted from ERA-Net Smart Grids Plus [1,2]).

1.1.1. Stakeholders Layer

In this layer stakeholders cover a diverse group of entities, ranging from individual end-users, communities, network operators, and aggregators to (local) governmental organizations. Stakeholders interact with smart grids through an interest or concern. Some features of smart grids, such as demand side management (DSM) and exchange of energy with other end users (peer-to-peer trading) involve individual end-user type stakeholders of the type of individual end users, whereas other features—such as a high penetration of renewable energy at a local level, as well as electric mobility [2,4–6]—involve stakeholders of a more organizational character such as network operators and governmental organizations.

1.1.2. Markets Layer

This layer comprises all the financial and business-related aspects of smart grids. Energy market structures, the micro-economics of energy technologies and energy billing belong to the markets layer. Also factors such as investments, net present value (NPV), levelized costs of electricity (LCoE), electricity tariffs and pricing mechanisms, fall into this category [7,8].

1.1.3. Technologies Layer

The technologies layer covers all the technological aspects of smart grids [9–12] related to energy technologies and information and communications technology (ICT), among which, but not exclusively, smart grid networks, distributed energy resources (DER) such as photovoltaic (PV) systems, wind turbines, micro-combined heat and power (μCHP) [13,14], energy storage systems [15], home energy management systems (HEMSs), DSM [16], demand shifting [17], demand and supply forecasting algorithms, electric vehicles (EVs), EV charging stations, stationary fuel cells, and hydrogen fuel cell electric vehicles (FCEVs) [18].

1.1.4. Flexibility

In addition to the three layers, special attention is given to 'flexibility' in this paper, as it is a main characteristic of future energy systems. It is defined as a "general concept of elasticity of resource deployment providing ancillary services for the grid stability and/or market optimisation" according to CENELEC [19]. In other words, electrical flexibility is the ability of a system to deploy its resources to respond to changes in net load, where the net load is defined as the remaining system load not served by variable generation [20]. Because the intermittent nature of renewable energy generation may threaten the stability of the overall system, smart grids can provide the flexibility needed for the correct operation of the energy system.

1.2. Aim

The main aim of this review was to emphasize the importance of each of the three layers and the current knowledge of each layer, to show the number of research activities in the different disciplines, and make an attempt to define the key elements of residential smart grids for sustainable energy and flexibility. To this end, a thorough review of the literature was performed. Compared to other established frameworks—such as the Transactive Energy [21], the Universal Smart Energy Framework (USEF) [22], and the Energy Flexibility Platform & Interface [23] —our review approaches energy transition at the residential level, using the theoretical three-layer model framework. We take into consideration each layer and their interrelations, suggesting cooperation among disciplines and parties, from their very beginnings and into the design phase. The main advantage of our approach is that we distinguish among the various disciplines in order to allocate knowledge and barriers for each layer in residential smart grid projects. This is a different approach than most of the stochastic or techno-economical models [24,25]. In this way, we aimed to decrease the uncertainties and increase active involvement at the residential level, making stakeholders and prosumers (consumers that also generate energy) part of the energy transition, and to stimulate feedback from end-users to designers. The research approach is presented in Section 2, with insights from the literature regarding the three layers then being presented (Section 3) and discussed (Section 4). The paper is summarized in the Conclusions (Section 5).

2. Research Approach

Our research consisted of reviewing journal papers, conference papers, reports and websites of interest to smart grids in the framework of technologies, markets, and stakeholders. The term 'smart grid' was used for the first time in 1966 [26]. Caution should be taken because at that time the term 'smart grid' was related to radio wave transmissions instead of electricity grids. The first official definition of a smart grid was approved by the US Congress in 2007, and signed into law in the same year [27]. According to a search in Scopus, which took place in November 2018, since 2007, there have been more than 100,000 papers published mentioning the term 'smart grid'. We found it most significant to consider articles that mentioned 'smart grid' in the abstract, title or keywords in our literature study. We found around 40,000 works, 92% of them being conference papers (64%) and articles (28%). The literature study is summarized in Figure 2, and pertains to the years and different layers proposed above.

Up to 2012, the number of papers was exponentially increasing for all areas of research into smart grids. Since 2012, the increasing annual number of publications on smart grid topics has led to a massive volume of more than 4000 publications per year, represented by a linear increase between 2012 and 2015. Around 5600 publications were published in 2016, which is very similar to 2017. At the beginning of November 2018, the number of publications in 2018 had already reached 5200. Although the year has not ended yet, we did include 2018 in our results (Figure 2). Based on affiliation, most of the publications originated from the EU (37.5%), the USA (22%), China (19%), and other countries (21%). The trends are similarly exponential before 2012 for almost all regions, and for all disciplines. After 2012, the number of publications stabilized around 1000 publications per year for the USA and China, although trends in EU countries vary considerably. The Joint Research Centre of the European Commission database and reports indicate that 953 smart grid projects have been funded in the EU since 2007, by 2900 different organizations, involving 5900 participants, and with investments of around €5 billion [28]. This diversity in projects and participants, gave the EU the lead in the number of scientific publications. For the USA, the number of projects announced on government websites was limited to only 119. As such, the related budget ($4.5 billion) indicates that projects funded in the USA were mostly larger-scale projects than in the EU [29]. For China, pilot projects were expected to be even more concentrated, as today there are 15 smart city pilots, with an expected $7.4 billion of investments by 2020 [30].

A further analysis was needed in order to categorize papers in terms of the three layers proposed of the framework, aimed at highlighting the trends in weight of research directed towards each layer. The boundaries between the disciplines are difficult to clarify. A classification based on which journal the publication appeared, and it led to relatively very small number of papers for the market (Business, Management, and Accounting: 2.6%) and stakeholders (Social Sciences Journals: 4.2%) layers. Therefore, we searched in Scopus using the keywords that we used to define the layers. Multidisciplinary studies were taken into consideration so publications dealing with than one layer were mentioned in both corresponding layers. Therefore, the sum of the publication for each layer is greater than the total number of papers. It appears that markets and stakeholder layers' tendencies are very close to each other. A possible explanation could be that many social scientific studies on users were critical of the market layer's scientific contributions or insights, and those papers thus represent critiques. Another explanation could be that user behavior, stakeholder investments and expectancies played a major role in the market layer, considering the dependency of energy price on demand. The technologies layer showed at least 50% more publications than the other layers regardless of the year of publication.

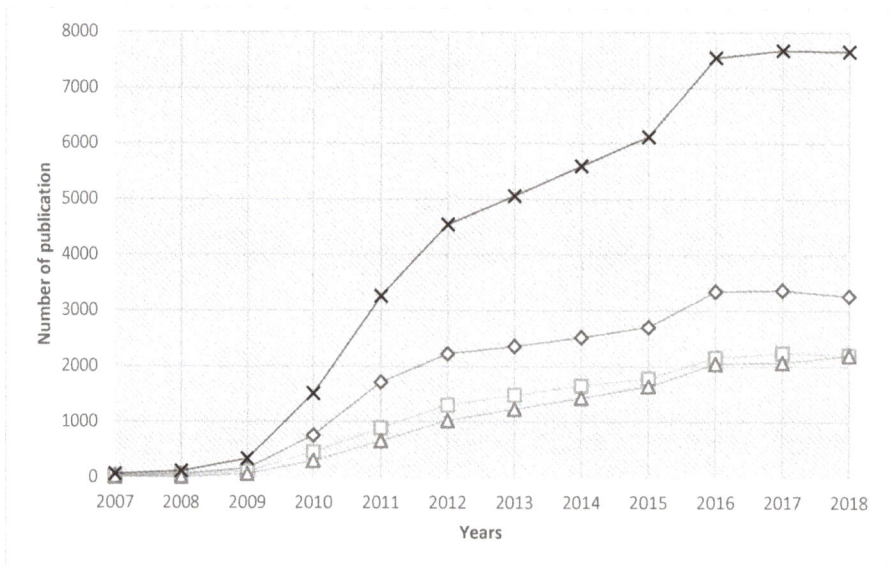

Figure 2. Yearly number of publications found in Scopus using the search term 'smart grid' and the three layers classification.

3. Findings for Each of the Three Layers

3.1. Stakeholders Layer

Looking at the stakeholders layer, the main focus is on residential customers and prosumers [31–33]. It was found that users, as they are described in the more recent research, not only have different labels—such as customers, consumers, prosumers and end users—but are also assumed to behave differently from each other [4,34–36]. The most prevalent type reported on in the literature is the consumer, who is supposed to adapt to new developments in the energy system, such as smart grids. The prosumers are seen as users who consume and (co-)produce energy, who are sometimes also seen as potential active players on energy markets through aggregators [37,38]. The prosumers' most important attribute is their 'proactiveness' inside the new energy system, which differs from passive consumers, who merely have to accept or adopt smart grids, and end users. Furthermore, based on the papers that focused on user experiences, with special attention paid to demand side management, it appears that the issue of the acceptance of smart grids is widely discussed. Only a few studies built on evaluations of real life experiences based on smart grid projects [39,40], whilst most studies focused on future scenarios and (online) surveys [32,40–42] (see report [43] for further details).

In particular, end-users' knowledge with regard to smart grids, is shown to both enhance acceptance and to create confusion [33,44,45]. The literature review shows that from current smart grid projects and the reference framework USEF, the following other main stakeholder groups can be distinguished in smart grid environments: business customers, aggregators, balancing responsible parties, balancing service providers (BSPs), suppliers, distribution system operators (DSOs), transmission system operators (TSOs), governments, and other regulators (Table 1) [8].

The role descriptions shown in Table 1 indicate stakeholders' roles and their interactions with the markets or technologies layers. Our study also shows that the future roles of other stakeholders, other than the residential customers, are foremost mentioned in policy reports with a special focus on the role of the DSOs, which are generally public organizations. Mostly, their concerns are reliability and equity among the residential energy consumers, and not the development of new markets. From these reports, it seems that uncertainties still exist with regards to market structures (monopoly versus competition), task delegation, the necessity of new (independent data handling) institutions, etc.; however, again, there is very little information about the exact roles of the stakeholders inside smart grids pilot projects, nor as facilitators of the renewable energy transition.

DSOs in the Netherlands play a major role in improving the reliability and robustness of the local grids in response to the supply of distributed renewable energy and expected peaks in demand for charging EVs. To support this statement, Dutch DSOs participated in 28 out of 31 projects with at least 15 households in each. In 12 of these pilots, they were part of the project consortium [43]. Regarding TSOs, their role is closely related to that of the DSOs, with the duties of developing and maintaining the transmission of electricity, and balancing supply and demand across different districts. They usually have a particular interest in new DER technologies as vehicle-to-grid (V2G) in order to provide local balance and supply to decrease transmission and avoid congestion. In the Netherlands, the TSO also pilots some of the smart grids with aggregators.

Table 1. Stakeholders in smart grids environments and descriptions of their roles (adapted from [8])

Stakeholder Group	Description
Residential customer/prosumer	A residential customer or utility business that produces electricity. Roof top PV installations and energy storage battery systems are examples of homeowner investments that allow people to do both consume and produce energy for use locally or to export during certain parts of the day or year.
Aggregator	A person or company combining two or more customers into a single purchasing unit in order to negotiate the purchase of electricity from retail electric providers, or the sale of electricity to these entities. Aggregators also combine smaller participants (as providers or customers or curtailment) to enable distributed resources to play in the larger markets.
Balancing responsible party (BRP)	A legal entity that manages a portfolio involving the demand and supply of electricity, and has a commitment to the system operator in a European Network of Transmission System Operators for Electricity (ENTSO-E) control zone to balance supply and demand in the managed portfolio on a Program Time Unit (PTU) basis according to energy programs.
Balancing service provider (BSP)	In the EU Internal Electricity Market, this is a market participant providing balancing services to its connecting transmission system operator (TSO), or in case of the TSO-BSP model, to its contracting TSO.
Supplier	A supplier provides energy to end customers, based on a contract. The energy can be from the supplier's own power plants or traded in relevant markets.
Distribution system operator (DSO)	The DSO is responsible for the safe and secure operation and management of the distribution system. DSOs are also responsible for the planning and development of the distribution system.
Transmission system operator (TSO)	A legal entity responsible for operating, developing, and maintaining the transmission system for a specific zone and, where apposite, its interconnections with other systems, and for guaranteeing the long-term ability of the system to meet reasonable demands for the transmission of electricity.
Government/Regulator	The regulator must strengthen competition and ensure that this does not compromise security of supply and sustainability. To act even-handedly in the interests of all market participants, regulators must be politically and financially independent.

3.2. Market Layer

From the existing literature, it can be concluded that several incentives for smart grid environments are present at the market level. Namely, aggregators can operate on the spot markets, i.e., by energy arbitrage, and on the balancing markets [8], however, some market barriers are present. For example, in many European countries, it is impossible for renewable electricity generation to operate on balancing markets, while this generation is very suitable to use for downward regulation [46]. Another current market inefficiency is the risk that current renewables generation incentive schemes (especially feed-in tariffs) decrease the value of newly installed renewables generation over time. Smart grids can address this by more efficiently matching supply and demand.

From a market perspective, one could argue that, when the shares of renewables in the grid increase to high levels, their inherent fluctuations would cause more volatile spot market prices and higher imbalance prices, thus providing higher incentives, and possibly business models, for smart solutions. On the other hand, one could also argue that, before that was the case, stakeholders would need to (and will) gain experience in these smart solutions because of the pivotal role that the electricity system plays in our society. Whether the current market model is already suitable for deploying smart grids thus remains therefore a matter for discussion in forthcoming years.

3.2.1. Pricing of Electricity and the EU Electricity Market

Concerning the pricing of electricity, in nodal pricing (or locational marginal pricing), which is incorporated in the electricity system of the USA, prices are set at different nodes in the system (places where supply and demand meet). In zonal pricing, as used in the EU electricity markets, prices are the same across the entire zone, not taking transmission limits into account. Therefore, the criticism of zonal pricing is that it does not stimulate the optimal placement of variable renewable electricity production [47,48]. For example, in Germany, much wind electricity production is located in the north, but the transmission line does not have the capacity to transfer this electricity to the south, resulting in congestion losses [47,49]. In 2015, 566 TWh was traded on the European Power Exchange (EPEX, including Germany, Austria, Luxembourg, France, the UK, the Netherlands, Belgium, Switzerland), while 59 TWh was traded on the intraday market, although the intraday market grew faster (26% versus 20%) [50].

There may be concerns about price and power fluctuations, regarding the market and technologies layers; however, flexible market scenarios, which control unwanted fluctuations, have recently been reported [51,52]. In addition, moderate fluctuations may potentially accelerate DER deployment, the design of various smart energy products and services, widen the international electricity networks' power transfer capacities and agreements as part of EPEX, meaning that one aspect viewed as unbeneficial for one layer, may come to be potentially beneficial for the future of smart grids. For such perspectives, the three-layer approach facilitates the identification of interactions between layers, and therefore permits more global and multidisciplinary approaches on smart grids and energy transitions.

3.2.2. Flexibility of the Market

In terms of the dynamics in the electricity supply system, and due to system stability reasons, demand and supply need to match at any instant in time; otherwise, the system is in imbalance. In liberalized electricity systems, market mechanisms implemented to maintain the balance are increasing. In short-term markets and balancing especially, flexibilities aggregated from end-users (customers or prosumers) provide an increasing potential for smart grids. Such flexibility can be used for several use cases, such as balancing, optimizing trading costs and minimizing costs from the imbalance settlement (e.g., caused by forecast errors in renewable electricity generation) or for the customer to increase their own consumption, whereby these use cases can be associated with different roles/actors.

When using the flexibility for one use case, however, several other actors may be influenced by this activation, either positively or negatively. Future markets in smart grid environments indicate an increased coordination need between several actors in regard to the integration of flexibility, and different flexibility use cases are depicted based on different roles, as shown in Figure 3.

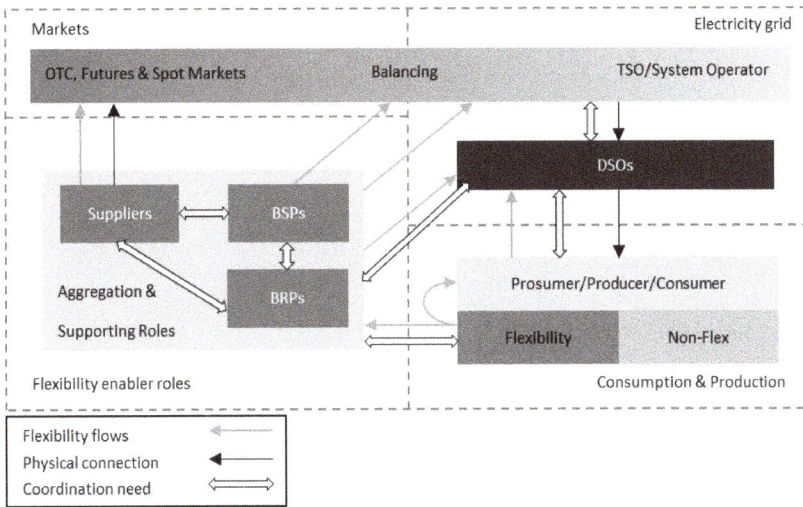

Figure 3. Stakeholder overview showing the flow of flexibility.

3.3. Technologies Layer

The technologies layer covers systems, technologies, and energy products and services that physically create a smart grid environment. In this section, the physical systems of smart grid will be analyzed from three perspectives; DER, demand side flexibility and resource side flexibility. These perspectives were to highlight smart grid flexibility, which is considered to be the key component for its integration into the electricity network [44].

3.3.1. Distributed Energy Resources

From the perspective of residential smart grids the following technologies can be distinguished: DER systems in the form of micro-generators and energy storage, smart appliances, smart meters, energy monitoring, and home automation (Table 2).

Table 2. Technologies layers in a smart energy system (adapted from [9,36]).

Category	Example
Distributed energy resource systems: micro-generators	Electricity: - Photovoltaic solar systems - Wind turbines Electricity and heat: - Micro cogeneration units - Fuel cells - Hybrid and fuel cell electric vehicles (FCEVs) - Solar heating and cooling
Distributed energy resource systems: energy storage	Electricity: - Batteries (household or neighborhood size) - Electrolyzers Heat: - In home hot water storage - Storage heaters - Shared storage in buildings or neighborhoods - Ground, aquifers, phase-change materials, thermochemical materials, etc.

Table 2. *Cont.*

Category	Example
Responsive appliances	- Electric vehicles (Battery) - Heat pumps - Air conditioners - Dish washers - Washing machines - Clothes dryers - Freezer/refrigerator - Battery operated home appliances robots (vacuum cleaners, kitchen appliances) - 3D printers, robot arms - Close-in boilers
Smart/digital meters	- Electricity meters (frequency ranges from seconds to day intervals) - Gas meters - Meters that allow for breakdown to appliance level (usually part of a monitoring and control system)
Energy monitoring and control systems	- Sensors and energy monitoring systems, ranging from household aggregate to breakdown to appliance level - Gas measurement, often combined with a smart thermostat
Home automation for smart energy use	- Energy services gateways - Apps - Steering of deferrable load (smart appliances) - Home automation and control - Internet of things - Smart plugs and smart battery chargers (lighting, USB grids, etc.)

With reference to the technologies layer, mainly capabilities regarding the flexibility of specific technologies or residential applications are of interest. In addition to the above, hydrogen as a storage medium has received much attention in the last few years because of the flexibility it can provide [53,54]. Electrolyzers can provide wind/solar peak shaving by splitting water with renewable energy sources electricity and producing hydrogen. This bulk energy storage process is known as power-to-gas. At peak demand, fuel cells can use hydrogen to quickly respond to the load demand. Hydrogen energy technologies are complementary to batteries, which can supply day-to-day electricity, whilst hydrogen can be used for long-term energy storage, particularly seasonal storage (summer-to-winter).

3.3.2. Demand Side Flexibility

Technical flexibility stresses two main aspects: controllability (e.g., on/off modes, shiftability, modulation, and so on) and characteristics (e.g., minimum and maximum power, reaction time, etc.). Apart from these general characterizations of resources, metrics can be introduced to specify a certain resource with respect to its capability, or 'characteristics'. In [55], different approaches for defining flexibility in terms of characteristics are discussed, in which the three main aspects identified are ramp magnitude, ramp frequency, and response time. Figure 4 shows a classification of the residual load for the stakeholders group 'prosumer', in terms of controllability. The base load represents the consumption in the prosumer premises, without any degree of freedom for flexibility or controllability. Loads providing controllability can be distinguished along with their ability to curtail, shift or store energy. As a concrete example, we can mention smart washing machines or dishwashers, which can provide a controllability, albeit within limits [56], to the smart grid for it to activate before the latest run time defined by the user [17]. In the case where a user programs the schedule of the appliance for economic reasons, the smart grid will optimize this according to the three layers: stakeholder (objective), market layer (price), technology (demand shifting).

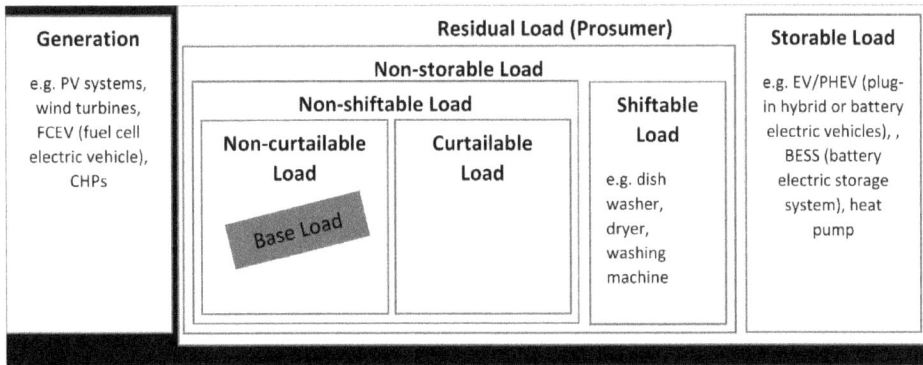

Generation	Residual Load (Prosumer)			Storable Load

Figure breakdown:
- **Generation**: e.g. PV systems, wind turbines, FCEV (fuel cell electric vehicle), CHPs
- **Residual Load (Prosumer)** — **Non-storable Load** — **Non-shiftable Load** — **Non-curtailable Load** (Base Load), **Curtailable Load**; **Shiftable Load** e.g. dish washer, dryer, washing machine
- **Storable Load**: e.g. EV/PHEV (plug-in hybrid or battery electric vehicles), , BESS (battery electric storage system), heat pump

Figure 4. Prosumer residual load in terms of controllability (adapted from [57]).

3.3.3. Resource-Side Flexibility

While generators and large resources can be described in terms of three main technical aspects, such as ramp magnitude, ramp frequency, and response time, smaller demand resources might need to be characterized using more details.

Based on individual technical parameters, the eligibility of the system types can be evaluated for smart energy product and service applications. The energy resources described in this section can be assigned following the categories of controllability, shown in Figure 4 at the beginning of this section. Regarding the characteristics of resources, Table 3 provides an overview of some these for individual systems.

Table 3. Characteristics of resources in terms of flexibility

Resource Type	Availability	Reaction Time	Duration
PV System	Depends on weather and time of day	seconds	Depends on weather and time of the day
Heat pumps	Fully available until temperature criterion of household is met	seconds	Fully available until temperature criterion of household is met
White goods and appliances	Customer dependent	seconds	Process dependent (non-interruptible)
Wet appliances	Customer dependent	seconds	Appliance dependent
Thermal storages	Fully available until temperature criterion is met	seconds	Limited by maximum battery capacity
Battery electric vehicles (EV)	Customer dependent	seconds	Limited by maximum battery capacity and customer
Fuel cell electric vehicles (FCEV)	Customer dependent	seconds	Limited by tank capacity, contrary to EV, tank fills as fast as 4 min

The availability and activation duration of all the investigated systems is somehow limited either by technical parameters or by the needs or influence of customers. In terms of power supply or demand, most systems range within typical power connection values of households in Europe (~17 kW). Depending on system configurations, EVs or PV systems may exceed this value significantly, which may lead to increasing needs for grid investments (e.g., transformers, power lines, etc.). It was shown that the investigated systems at the residential level provide a different potential for flexibility applications. Whilst existing systems—such as PV systems, heat pumps, or appliances—provide limited controllability (except for downward regulation), the introduction of stationary battery energy storage systems in particular can enable full flexibility for local optimization or ancillary services.

4. Discussion and Perspectives

Although smart grids are still in an early stage of development, in recent years, societal implementation has gained momentum through the deployment of smart meters and small and medium scale smart grid pilots [44,58–60]. The transition to smart grids would create electricity systems that would enable consumers to make informed and empowered energy-related choices, promoting personal behavioral changes [39,61]. In this regard, evaluative studies and reports, such as [62–65] have highlighted the relevance of end-users in smart grid deployment. Nowadays, statistically significant data concerning social factors in stakeholder issues are few compared to the number of papers on technologies and market layers analyses [1]. Social-acceptance, in regard to product adaptation, appears to be a slow process, nevertheless, it is one of the key factors in the fast deployment of smart grids into the actual grid [39]. The continuous interactions and co-evolution of the three layers will define the smart grid environment, and the more interaction there is, the more rapid the transition will be to a fully smart grid. Our study aimed to highlight this multi-layer interaction to enable a wider point of view and practical solutions.

The first crucial point for social implementation is the selection bias in monitoring surveys caused by the so-called Hawthorne effect (see [66]). For instance, one of the world's largest datasets of appliances, Pecan Street [67], is sharing the circuit-level residential electricity data in the USA intended for research purposes. The data has shown that the monitored residences are consuming far less electricity (33–60%) compared than the means of that specific region's other network users, probably due to being a volunteer or being conscious that they are participating in a pilot experiment. Active participation of the individuals is crucial for efficient integration and energy resilience, therefore the design of smart products and services is crucial to maintaining their motivation to be energy efficient. In the stakeholders layer, detailed consumer profiles analyses are needed in order to point out such important facts, which will guide the technologies and market layers on structural issues in the electricity network. Treating raw data according to only the market and technologies layers, without considering end-users and stakeholders, is already inducing unneglectable simulation uncertainties on an hourly basis, and the high prediction errors may induce significant organizational and structural errors. The data purged of bias will help us to focus on specific smart energy products and services, their research and development [68].

Pecan Street's recent data analyses from 12,083 monitored residences showed that 50% of the electricity used was related to air conditioning units, water heaters, and refrigerators [69]. Such applications could be shiftable loads if thermal isolation was feasible and sufficient, which again underlines the importance of load flexibility management. The main barrier would be the initial costs of these zero energy households, additionally equipping them with devices such as heat pumps, which in the EU are now supported by policy. Creating facilities and smart energy products, depending on end-users' surveys about these devices and comfort expectancies, would be a cost-effective solution. Moreover, users' knowledge with regard to smart grids (such as an abundance of feedback information) has been shown to both enhance acceptance and to create confusion, and in contexts, people prefer demand load control (the ability of energy suppliers to control user consumption) while they reject it in others.

For EU countries, the risk of residential peak demand is high and similar for countries such as Italy, Norway, and Germany, for average households, while on the other hand, Bulgaria and the UK differ completely [70]. Global warming also plays a role in the trend of peak demand. In 2014, France had the lowest peak demand on record since 2004 [71]. Other EU countries' consumption profiles and peak demand characteristics remain to be validated for the smart grids from ongoing EU pilot projects [1]. At the regional level, zone policies and network characteristics vary. For instance, for the

Netherlands, the barriers to fast deployment of smart grids are mostly uncertainty of the benefits, with only about 11.1% of the network contains renewable energy generation in 2015 [72], while a target of 14% renewables by 2020 in the Netherlands has been stated [73]. Meanwhile, Austria has a target of 34% renewable energy by 2020, and 100% self-sufficiency in energy by 2050 [74]. The main reason for this is the use of hydroelectricity as a huge energy reservoir to flatten the demand curve, which improves the renewable energy cost effectiveness by erasing the need for residential batteries. Austria is actually marketing itself as the 'battery of Europe'. Transnational collaborations and knowledge sharing initiatives between these types of countries are in progress, in order to learn from favorable conditions and to consider weak points, so as provide more solid initiatives [1]. For instance, in the Netherlands, there are some investment plans in operation until 2025, as the green hydrogen economy in north, which might be a solution for the electricity grid capacity and flexibility problems in a cost effective way, through combining with other renewable/sustainable energy sources [75]. The renewable energy seasonal surplus could be converted and stored as hydrogen, transferred across the country by the existing gas network (after minor conversion), and could be stored in salt caverns to provide seasonal flexibility, and to balance regional pros and cons, salt caverns and the dramatic decrease in renewable energy production during winter.

Regarding energy efficiency and sustainable energy usage by EU citizens, the three-layer model usage will keep their motivations in pace, as feeling part of the energy transition. Smart products and services would give them the possibility to become greener or accomplish their economic objectives giving them a certain degree of freedom, meanwhile also providing the ability to the smart grid to optimize the shiftable load whenever available. Techno-economical approaches, or fully automated demand load control, or massive deployment of smart grids, would certainly bring resilience and efficiency from a technical and organizational perspective, which is of course a necessity [24,25]. However, in the residential sector, if not combined with the stakeholders layer, these approaches might fail to make citizens become more energy resilient or sustainable. Furthermore, because they might not feel part of the energy transition, the usage performance of new technology devices might drop. Or even worse, it paradoxically increase citizens' consumption, as they may tend to consume more due to disempowerment resulting from the automated processes. Automated demand load would certainly reduce the peak demand, by risking this to be perceived as a real constraint for the residential sector. Even used to a moderate extent, if end-user perspectives are not taken into account, they will greatly reduce the acceptance of smart grids. Our three-layer model aims to highlight such interactions and the possible consequences, where the solution should come from all layers, not only from a top-down approach. The main barrier to smart grid deployments is the lack of multidisciplinary considerations for the residential sector where uncertainties are vast and objectives, consumption and production patterns, geographic attributes, local authority aims and policies are not identical.

5. Conclusions

Our study aimed to present a three-layer model (stakeholders, markets, technologies) for the assessment of residential smart grids. In this way, knowledge about the actual performance of residential smart grids can be collected and evaluated within its framework. The use of the three-layer approach increases awareness of the multidisciplinary aspects of the problem, which most of the recent and extensive technical literature reviews point out [11].

We defined and discussed each layer in terms of recent issues from different perspectives. Uncertainties still exist with regard to market structures, task delegation, the necessity for new (data-handling) institutions, etc. Issues resulting from human factors and the stakeholders layers, comfort expectancies, the aims and requirements of smart grid users are still unknown parameters, however they are deemed to be crucial in order to fit the market and demand energy management systems in smart grids. If the goal of some consumers is to be more sustainable, and for others to relinquish any comforts, but only to sell some of their local energy production, then the conflict of interest has to be analyzed well considering the entire grid. Consuming only renewable energies and storing them may not be the greenest way, as batteries also have notable ecological impacts. Bottom-up end user and stakeholder capacities and demands must be analyzed in conjunction with statistical consumption graphs. Also, a portfolio of users identified for minimizing bias, should be considered in order to discern structural and pricing issues in smart grids. Modellers admit the necessity of including many parameters, especially stakeholder behavioral or characterization parameters [54]. Social acceptance and practices also have to be considered closely and more data is needed to be more statistically significant, in order to define to what extent flexibility may play a role [28].

Funding: This research has received funding from the European Union's Horizon 2020 research and innovation programme under the ERA-Net Smart Grids plus, grant number 646039, from the Netherlands Organisation for Scientific Research (NWO) and from BMVIT/BMWFW under the Energy der Zukunft programme.

Acknowledgments: Our project has received funding in the framework of the joint programming initiative ERA-Net Smart Grids Plus, with support from the European Union's Horizon 2020 research and innovation programme. We would like to thank Esin Gültekin for her contributions on literature studies. Furthermore, we would like to acknowledge all participants in the smart grid pilots (in the Netherlands) involved in this study for their willingness to share their data, experiences, and knowledge with the researchers.

Conflicts of Interest: The authors declare no conflict of interest.

Disclaimer: The content and views expressed in this material are those of the authors and do not necessarily reflect the views or opinion of the ERA-Net SG+ initiative. Any reference given does not necessarily imply the endorsement by ERA-Net SG+.

References

1. Reinders, A.; de Respinis, M.; van Loon, J.; Stekelenburg, A.; Bliek, F.; Schram, W.; van Sark, W.; Esteri, T.; Uebermasser, S.; Lehfuss, F.; et al. Co-evolution of smart energy products and services: A novel approach towards smart grids. In Proceedings of the 2016 Asian Conference on Energy, Power and Transportation Electrification, Singapore, 25–27 October 2016; pp. 1–6.

2. ERA-Net Smart Grids Plus. *European Research Area Network – Smart Grids Plus*; ERA NET: Amsterdam, The Netherlands, 2017; pp. 5–8.

3. International Energy Agency. *World Energy Outlook*; Organization for Economic Co-Operation and Development (OECD): Paris, France, 2011; ISBN 978-92-64-12413-4.

4. Geelen, D.; Vos-Vlamings, M.; Filippidou, F.; van den Noort, A.; van Grootel, M.; Moll, H.; Reinders, A.; Keyson, D. An end-user perspective on smart home energy systems in the PowerMatching City demonstration project. In Proceedings of the IEEE Innovative Smart Grid Technologies Europe (PES ISGT) Europe 2013, Copenhagen, Denmark, 6–9 October 2013; pp. 1–5.

5. Verhoef, L.; Graamans, L.; Gioutsos, D.; van Wijk, A.; Geraedts, J.; Hellinga, C. ShowHow: A Flexible, Structured Approach to Commit University Stakeholders to Sustainable Development. In *Handbook of Theory and Practice of Sustainable Development in Higher Education*; Leal Filho, W., Azeiteiro, U.M., Alves, F., Molthan-Hill, P., Eds.; Springer: Cham, Switzerland, 2017; pp. 491–508. ISBN 978-3-319-47876-0.

6. Griffiths, J.; Maggs, H.; George, E. *Stakeholder Involvement*; World Health Organization (WHO): Geneva, Switzerland, 2007.

7. Metz, D. Economic Evaluation of Energy Storage Systems and Their Impact on Electricity Markets in a Smart-grid Context. Ph.D. Thesis, University of Porto, Porto, Portugal, 2017.

8. Universal Smart Energy Framework. *USEF: The Framework Explained*; USEF Foundation: Arnhem, The Netherlands, 2018.

9. Van Wijk, A.; Verhoef, L. *Our Car as Power Plant*; IOS Press: Amsterdam, The Netherlands, 2014; ISBN 978-1-61499-377-3.

10. van Wijk, A.; van der Roest, E.; Boere, J. *Solar Power to the People*; IOS Press BV: Amsterdam, The Netherlands, 2017; ISBN 978-1-61499-832-7.

11. Nosratabadi, S.M.; Hooshmand, R.-A.; Gholipour, E. A comprehensive review on microgrid and virtual power plant concepts employed for distributed energy resources scheduling in power systems. *Renew. Sustain. Energy Rev.* **2017**, *67*, 341–363. [CrossRef]

12. Oldenbroek, V.; Verhoef, L.A.; van Wijk, A.J.M. Fuel cell electric vehicle as a power plant: Fully renewable integrated transport and energy system design and analysis for smart city areas. *Int. J. Hydrogen Energy* **2017**, *42*, 8166–8196. [CrossRef]

13. Gercek, C. Evaluation of heat pumps for balancing grids in combination with solar energy production: A Dutch Case Study. In Proceedings of the Solar Integration Workshop 2018, Stockholm, Sweden, 15–19 October 2018.

14. Gercek, C.; Reinders, A. Photovoltaic Energy Integration: A Case Study on Residential Smart Grids Pilots in The Netherlands. In Proceedings of the 35th European Photovoltaic Solar Energy Conference and Exhibition (EU PVSEC), Brussels, Belgium, 24–28 September 2018.

15. Schram, W.L.; Lampropoulos, I.; van Sark, W.G.J.H.M. Photovoltaic systems coupled with batteries that are optimally sized for household self-consumption: Assessment of peak shaving potential. *Appl. Energy* **2018**, *223*, 69–81. [CrossRef]

16. Weck, M.H.J.; van Hooff, J.; van Sark, W.G.J.H.M. Review of barriers to the introduction of residential demand response: A case study in the Netherlands: Barriers to residential demand response in smart grids. *Int. J. Energy Res.* **2017**, *41*, 790–816. [CrossRef]

17. Gercek, C.; Reinders, A. Balancing Renewable Energy Sources in Electricity Grids the Netherlands—How Residential Smart Grids can Contribute to Flexibility of Grids. In Proceedings of the DIT-ESEIA Conference on Smart Energy Systems in Cities and Regions, Dublin, Ireland, 10–12 April 2018.

18. Robledo, C.B.; Oldenbroek, V.; Abbruzzese, F.; van Wijk, A.J.M. Integrating a hydrogen fuel cell electric vehicle with vehicle-to-grid technology, photovoltaic power and a residential building. *Appl. Energy* **2018**, *215*, 615–629. [CrossRef]

19. CEN-CENELEC-ETSI Smart Grid Coordination Group. Smart Grid Reference Architecture. Available online: https://ec.europa.eu/energy/sites/ener/files/documents/xpert_group1_reference_architecture.pdf (accessed on 1 November 2018).

20. Lannoye, E.; Flynn, D.; O'Malley, M. Evaluation of Power System Flexibility. *IEEE Trans. Power Syst.* **2012**, *27*, 922–931. [CrossRef]

21. Chen, S.; Liu, C.-C. From demand response to transactive energy: State of the art. *J. Mod. Power Syst. Clean Energy* **2017**, *5*, 10–19. [CrossRef]

22. Universal Smart Energy Framework. *USEF: The Framework Specifications*; USEF Foundation: Arnhem, The Netherlands, 2018.

23. Flexiblepower Alliance Network (FAN). *Energy Flexibility Platform & Interface*; FAN: Delft, The Netherlands, 2018.

24. Marzband, M.; Fouladfar, M.H.; Akorede, M.F.; Lightbody, G.; Pouresmaeil, E. Framework for smart transactive energy in home-microgrids considering coalition formation and demand side management. *Sustain. Cities Soc.* **2018**, *40*, 136–154. [CrossRef]

25. Marzband, M.; Azarinejadian, F.; Savaghebi, M.; Pouresmaeil, E.; Guerrero, J.M.; Lightbody, G. Smart transactive energy framework in grid-connected multiple home microgrids under independent and coalition operations. *Renew. Energy* **2018**, *126*, 95–106. [CrossRef]

26. Darlington, S.; Felsen, L.B.; Siegel, K.M.; Deschamps, G.; Hansen, R.C.; Ishimaru, A.; Keller, J.B.; King, R.W.P.; Marcuvitz, N.; Senior, T.B.A.; et al. U.S.A. National Assembly, Committee Report, Fifteenth URSI General Munich, September 1966: Commission 6, Radio Waves and Transmission of Information; Progress. in Radio Waves and Transmission of. *Radio Sci.* **1966**, *1*, 1371–1379. [CrossRef]

27. US Government. *Energy Independence and Security Act*; 110th United States Congress, Public law 110-140; U.S. Government Printing Office: Washington, DC, USA, 2007.

28. Gangale, F.; Vasiljevska, J.; Mengolini, A.; Fulli, G. *Smart Grid Projects Outlook 2017: Facts, Figures and Trends in Europe*; Joint Research Centre: Brussels, Belgium, 2017.

29. Office of Electric Delivery and Energy Reliability for the SGIG. *Recovery Act Smart Grid Document Collection; Key Documents from DOE's Recovery Act Smart Grid Investment Grant and Demonstrations Programs*; US Department of Energy: Washington, DC, USA, 2016.

30. JUCCE. Smart Grid in China. Available online: https://www.juccce.org/smartgrid (accessed on 16 November 2018).

31. Michaels, L.; Parag, Y. Motivations and barriers to integrating 'prosuming' services into the future decentralized electricity grid: Findings from Israel. *Energy Res. Soc. Sci.* **2016**, *21*, 70–83. [CrossRef]

32. Fell, M.J.; Shipworth, D.; Huebner, G.M.; Elwell, C.A. Public acceptability of domestic demand-side response in Great Britain: The role of automation and direct load control. *Energy Res. Soc. Sci.* **2015**, *9*, 72–84. [CrossRef]

33. Horne, C.; Darras, B.; Bean, E.; Srivastava, A.; Frickel, S. Privacy, technology, and norms: The case of smart meters. *Soc. Sci. Res.* **2015**, *51*, 64–76. [CrossRef] [PubMed]

34. Van Vliet, B.; Chappells, H.; Shove, E. *Infrastructures of Consumption: Environmental Innovation in the Utility Industries*; Earthscan: London, UK; Sterling, VA, USA, 2005; ISBN 978-1-85383-996-2.

35. Goulden, M.; Bedwell, B.; Rennick-Egglestone, S.; Rodden, T.; Spence, A. Smart grids, smart users? The role of the user in demand side management. *Energy Res. Soc. Sci.* **2014**, *2*, 21–29. [CrossRef]

36. Geelen, D.V. *Empowering End-Users in the Energy Transition: An Exploration of Products and Services to Support Changes in Household Energy Management*; TU Delft: Delft, The Netherlands, 2014.

37. Naus, J.; Spaargaren, G.; van Vliet, B.J.M.; van der Horst, H.M. Smart grids, information flows and emerging domestic energy practices. *Energy Policy* **2014**, *68*, 436–446. [CrossRef]

38. Geelen, D.; Scheepens, A.; Kobus, C.; Obinna, U.; Mugge, R.; Schoormans, J.; Reinders, A. Smart energy households' pilot projects in The Netherlands with a design-driven approach. In Proceedings of the 2013 4th IEEE/PES Innovative Smart Grid Technologies Europe, ISGT Europe 2013, Lyngby, Denmark, 6–9 October 2013; pp. 1–5.

39. Smale, R.; van Vliet, B.; Spaargaren, G. When social practices meet smart grids: Flexibility, grid management, and domestic consumption in The Netherlands. *Energy Res. Soc. Sci.* **2017**, *34*, 132–140. [CrossRef]

40. Raimi, K.T.; Carrico, A.R. Understanding and beliefs about smart energy technology. *Energy Res. Soc. Sci.* **2016**, *12*, 68–74. [CrossRef]

41. Buchanan, K.; Banks, N.; Preston, I.; Russo, R. The British public's perception of the UK smart metering initiative: Threats and opportunities. *Energy Policy* **2016**, *91*, 87–97. [CrossRef]

42. Döbelt, S.; Jung, M.; Busch, M.; Tscheligi, M. Consumers' privacy concerns and implications for a privacy preserving Smart Grid architecture—Results of an Austrian study. *Energy Res. Soc. Sci.* **2015**, *9*, 137–145. [CrossRef]

43. Reinders, A.; Hassewend, B.; Obinnna, U.; Markocic, E.; de Respinis, M.; Schram, W.; van Sark, W.; Gultekin, E.; van Mierlo, B.; van Wijk, A.; et al. *Literature Study on Existing Smart Grids Experiences*; University of Twente: Enschede, The Netherlands, 2018; pp. 27–42.

44. Verbong, G.P.; Beemsterboer, S.; Sengers, F. Smart grids or smart users? Involving users in developing a low carbon electricity economy. *Energy Policy* **2013**, *52*, 117–125. [CrossRef]

45. Lopes, M.A.; Antunes, C.H.; Janda, K.B.; Peixoto, P.; Martins, N. The potential of energy behaviours in a smart (er) grid: Policy implications from a Portuguese exploratory study. *Energy Policy* **2016**, *90*, 233–245. [CrossRef]

46. Hu, J.; Harmsen, R.; Crijns-Graus, W.; Worrell, E.; van den Broek, M. Identifying barriers to large-scale integration of variable renewable electricity into the electricity market: A literature review of market design. *Renew. Sustain. Energy Rev.* **2018**, *81*, 2181–2195. [CrossRef]

47. Neuhoff, K.; Barquin, J.; Bialek, J.W.; Boyd, R.; Dent, C.J.; Echavarren, F.; Grau, T.; von Hirschhausen, C.; Hobbs, B.F.; Kunz, F. Renewable electric energy integration: Quantifying the value of design of markets for international transmission capacity. *Energy Econ.* **2013**, *40*, 760–772. [CrossRef]

48. Wang, Q.; Zhang, C.; Ding, Y.; Xydis, G.; Wang, J.; Østergaard, J. Review of real-time electricity markets for integrating distributed energy resources and demand response. *Appl. Energy* **2015**, *138*, 695–706. [CrossRef]

49. Scharff, R. *Design of Electricity Markets for Efficient Balancing of Wind Power Generation*; KTH Royal Institute of Technology: Stockholm, Sweden, 2015.

50. European Power Exchange. *EPEX SPOT Reaches in 2015 the Highest Spot Power Exchange Volume Ever*; European Power Exchange: Paris, France, 2016.

51. Qin, J.; Ma, Q.; Shi, Y.; Wang, L. Recent Advances in Consensus of Multi-Agent Systems: A Brief Survey. *IEEE Trans. Ind. Electron.* **2017**, *64*, 4972–4983. [CrossRef]

52. Jain, R.K.; Qin, J.; Rajagopal, R. Data-driven planning of distributed energy resources amidst socio-technical complexities. *Nat. Energy* **2017**, *2*, 17112. [CrossRef]

53. Valverde, L.; Rosa, F.; Bordons, C.; Guerra, J. Energy Management Strategies in hydrogen Smart-Grids: A laboratory experience. *Int. J. Hydrogen Energy* **2016**, *41*, 13715–13725. [CrossRef]

54. Patel, P.; Jahnke, F.; Lipp, L.; Abdallah, T.; Josefik, N.; Williams, M.; Garland, N. Fuel Cells and Hydrogen for Smart Grid. In Proceedings of the 2010 Fuel Cell Seminar & Exposition, San Antonio, TX, USA, 18–21 October 2011; pp. 305–313.

55. Lund, P.D.; Lindgren, J.; Mikkola, J.; Salpakari, J. Review of energy system flexibility measures to enable high levels of variable renewable electricity. *Renew. Sustain. Energy Rev.* **2015**, *45*, 785–807. [CrossRef]

56. Staats, M.R.; de Boer-Meulman, P.D.M.; van Sark, W.G.J.H.M. Experimental determination of demand side management potential of wet appliances in the Netherlands. *Sustain. Energy Grids Netw.* **2017**, *9*, 80–94. [CrossRef]

57. He, X.; Hancher, L.; Azevedo, I.; Keyaerts, N.; Meeus, L.; Glachant, J.-M. *Shift, Not Drift: Towards Active Demand Response and Beyond*; European University Institute (EUI): Florence, Italy, 2013.

58. Stephens, J.C.; Wilson, E.J.; Peterson, T.R.; Meadowcroft, J. Getting smart? climate change and the electric grid. *Challenges* **2013**, *4*, 201–216. [CrossRef]

59. Naus, J.; van Vliet, B.J.; Hendriksen, A. Households as change agents in a Dutch smart energy transition: On power, privacy and participation. *Energy Res. Soc. Sci.* **2015**, *9*, 125–136. [CrossRef]

60. Wolsink, M. The research agenda on social acceptance of distributed generation in smart grids: Renewable as common pool resources. *Renew. Sustain. Energy Rev.* **2012**, *16*, 822–835. [CrossRef]

61. DeWaters, J.E.; Powers, S.E. Energy literacy of secondary students in New York State (USA): A measure of knowledge, affect, and behavior. *Energy Policy* **2011**, *39*, 1699–1710. [CrossRef]

62. European Consumer Markets Evaluation Consortium. *The Functioning of Retail Electricity Markets for Consumers in the European Union*; European Commission: Brussel, Belgium, 2010.

63. *EU-Commission Energy 2020: A Strategy for Competitive, Sustainable and Secure Energy*; No. 639; Publications Office of the European Union: Luxembourg, 2010.

64. European Technology Platform. *SmartGrids Strategic Deployment Document for Europe's Electricity Networks of the Future*; European Technology Platform SmartGrids: Brussels, Belgium, 2008.

65. International Energy Agency. *World Energy Outlook*; Organization for Economic Co-operation and Development (OECD): Paris, France, 2017; ISBN 978-92-64-28205-6.

66. Schwartz, D.; Fischhoff, B.; Krishnamurti, T.; Sowell, F. The Hawthorne effect and energy awareness. *Proc. Natl. Acad. Sci. USA* **2013**, *110*, 15242–15246. [CrossRef] [PubMed]

67. Pecan Street. *Pecan Street Online Database*; Pecan Street: Austin, TX, USA, 2016.

68. Glasgo, B.; Hendrickson, C.; Azevedo, I.L. Assessing the value of information in residential building simulation: Comparing simulated and actual building loads at the circuit level. *Appl. Energy* **2017**, *203*, 348–363. [CrossRef]

69. Glasgo, B.; Hendrickson, C.; Azevedo, I.M.L. Using advanced metering infrastructure to characterize residential energy use. *Electr. J.* **2017**, *30*, 64–70. [CrossRef]

70. Torriti, J. The Risk of Residential Peak Electricity Demand: A Comparison of Five European Countries. *Energies* **2017**, *10*, 385. [CrossRef]

71. Réseau de Transport d'Electricité. *2014 Annual Electricity Report*; Réseau de Transport d'Electricité: Paris, France, 2015.

72. Eurostat. *Your Key to European Statistics*; Eurostat: Luxembourg, 2015.

73. Ministry of Economic Affairs and Climate Policy. *Dutch Goals with EU*; Ministry of Economic Affairs and Climate Policy: The Hague, The Netherlands, 2017.

74. Federal Ministry of Economy, Family and Youth, Energy Strategy Austria. *Energy Strategy Austria*; Federal Ministry of Economy, Family and Youth, Energy Strategy Austria: Vienna, Austria, 2017.

75. Northern Innovation Board. *The Green Hydrogen Economy in the Northern Netherlands*; Northern Innovation Board: Groningen, The Netherlands, 2017.

applied sciences

MDPI

Article

A Comparison of Households' Energy Balance in Residential Smart Grid Pilots in the Netherlands

Cihan Gercek [1,*], Wouter Schram [2], Ioannis Lampropoulos [2], Wilfried van Sark [2] and Angèle Reinders [1,3]

[1] Department of Design, Production and Management, Faculty of Engineering Technology, University of Twente, P.O. Box 217, 7500 AE Enschede, The Netherlands

[2] Copernicus Institute of Sustainable Development, Utrecht University, Heidelberglaan 2, 3584 CS Utrecht, The Netherlands

[3] Energy Technology Group at Mechanical Engineering, Eindhoven University of Technology, P.O. Box 513, 5600 MB Eindhoven, The Netherlands

* Correspondence: c.gercek@utwente.nl; Tel.: +31-534897875

Received: 9 April 2019; Accepted: 19 July 2019; Published: 25 July 2019

Abstract: This paper presents an analysis that frames the impact of various smart grid technologies, with an objective to provide a transparent framework for residential smart grid demonstration projects based on predefined and clearly formulated key performance indicators. The analysis inspects measured energy data of 217 households from three smart grid pilot projects in the Netherlands and a public dataset with smart meter data from 70 households as a reference. The datasets were evaluated for one year and compared to provide insights on technologies and other differences based on seven key performance indicators, giving a comprehensive overview: monthly electricity consumption (100–600 kWh) and production (4–200 kWh); annually imported (3.1–4.5 MWh) and exported (0.2–1 MWh) electricity; residual load; peak of imported (4.8–6.8 kW) and exported (0.3–2.2 kW) electricity; import simultaneity (20–70.5%); feed in simultaneity (75–89%); self-sufficiency (18–20%); and self-consumption (50–70%). It was found that the electrification of heating systems in buildings by using heat pumps leads to an increase of annual electricity consumption and peak loads of approximately 30% compared to the average Dutch households without heat pumps. Moreover, these peaks have a high degree of simultaneity. To increase both the self-sufficiency and self-consumption of households, further investigations will be required to optimize smart grid systems.

Keywords: smart grids; flexibility; photovoltaic; heat pumps; consumption patterns; self-consumption; self-sufficiency; energy system analysis; load duration curve

1. Introduction

Energy transitions are gaining momentum with the digitalization of power systems and the use of smart grid technologies, which enable various possibilities to efficiently integrate renewable energy sources within the residential sector. In 2011, the European Commission acknowledged the importance of promoting smart grids and smart metering throughout the EU in their communication "Smart Grids: from innovation to deployment" [1]. Various benefits were highlighted: empowering consumers to directly control and manage consumption patterns [2,3], enabling time-dependent electricity prices, providing more cost-efficient energy use, enhancing security of the grid [4,5] enabling integration of renewable energy [6] and electric vehicles (EV) [7], boosting future competitiveness and technological leadership, and providing a platform for the development of innovative energy services from individual users and through aggregator companies [8–10].

The translation of these theoretical benefits towards practical applications has led to the implementation of many smart grid pilots within the EU [11]. According to the records of the

European Joint Research Center (JRC), a total of 527 smart grid pilots have been implemented in Europe, mostly categorized as Demonstration projects and Research and Development (R&D) projects. The total budget for these projects is 360 billion euro, with almost two third of the funds attributed to demonstration projects [12]. Another important parameter is the density of the investments. Namely, the Netherlands is cited as a high density spot in many reports [11,13]. The Dutch government has put emphasis on the importance of smart grid pilots [14–16] and integration of renewable energies [17,18]. Thanks to these policies, many differently configured energy systems have been implemented and tested in several residential smart grid pilots and demonstration projects, which are further elaborated in Section 2. In a conventional power grid, electricity consumption patterns of residential households(HH.) depend on geography, socio-economic factors of the households, and especially the number of residents per household, which we refer to as household size [19,20]. In a residential smart grid, the net electricity that is imported from the grid (called import in this paper) is not equal anymore to the consumption, and varies depending on additional local power production by distributed energy resources (DER) coupled with home energy management software (HEMS) [21]. This mainly entails rooftop photovoltaic (PV) systems [22], but also other DER as micro-combined heat and power (μ-CHP) [23]. Out of 84 projects executed in the Netherlands, 31 residential smart grid demonstration pilots involved more than 15 households per pilot and are well documented [24]. However, little information is available about the functioning of these new energy systems and their overall performances in combination with each other, as most studies tend to focus only on one aspect of electric load, which still contains a significant amount of valuable knowledge [25]. Therefore, this study analyzes four datasets in further detail, focusing on seven key performance indicators: (1) monthly electricity consumption and production; (2) annually imported and exported electricity; (3) residual load; (4) peak of imported and exported electricity; (5) simultaneity in import and feed-in; (6) and self-sufficiency and (7) self-consumption.

These key performance indicators (KPI) are related to the performance of energy technologies [26] and to the end-users (prosumers) in residential smart grid pilots [27,28]. These key energy performance indicators can be used to determine and quantify various aspects, namely, seasonal effects, extreme conditions during peak times and necessary import capacity and the degree of integration of renewable energy sources [26,29,30]. The annually imported electricity gives an indication of energy efficiency and provides insight into how much electricity must be generated. Residual load is the load that the grid operators observe, so grid load is important for matters, such as voltage fluctuations and the sizing of connections, cables, and transformers, related to the design of grids. Therefore, residual load is employed as an important parameter in EU reports, such as [31]. The same holds for peak import and export of electricity, which are, respectively, the maximum (positive) and minimum (negative, i.e., electricity feed-in) values of the residual load. These are essential metrics for dimensioning at the household level. Cables and transformers are not dimensioned by summing all the maximum import and export values of individual households, but by the aggregated values of multiple households. Therefore, on an aggregated level (e.g., a cable and a transformer), the simultaneity of both electricity import and PV feed-in are essential metrics [32]. Next, self-sufficiency quantifies how well a household can rely on its renewable energy production and independently operate from the local grid, and self-consumption figures are relevant for sizing distributed energy systems with respect to household demand [33–35].

The aforementioned indicators are often studied individually, whereas this study presents a more comprehensive overview. Adding to this novelty, normally only one data set is studied [36], while this study compares multiple data sets, increasing the robustness of the results. Furthermore, in this research, a distinction is made between the aforementioned indicators at a household level [35] and a community level. The reliability of the grid depends significantly on the aggregated level instead of the individual level [31,37,38].

In this paper, we analyze and compare the monitored time series of energy data from three different residential smart grid pilot projects and one set of smart meter data, namely, the Power Matching City (22) in Groningen [39–41], the Jouw Energie Moment (117) in Breda [42], and the Rendement voor

Iedereen (79) in Amersfoort [43]. Further, Zonnedael (70), an open access dataset of smart meter data provided by a Dutch distributed system operator (DSO) [44], is used as a reference and for comparison purposes. A total of 287 households were analyzed in this study. In Section 2, the pilots, the employed technologies, the data format, and the processing methods are described. In Section 3, the results are presented, and the pilots are compared according to seven predefined key performance indicators (KPIs). The paper is concluded with a discussion and recommendations for future studies.

2. Pilot Descriptions, Data and Processing Methods

In this section, the input for the data analysis of residential smart grid pilots and the analytical methods which are the basis of our algorithms in MatLab© are elucidated. In Section 2.1, we provide information about the smart grid pilots, and in Section 2.2, we present the types of data that were analyzed. In Section 2.3, we elaborate on the methods used for the data analysis.

2.1. Pilot Descriptions

Prior to the data analysis, in this section, we provide brief descriptions for the investigated pilots and the technologies employed within these pilots. Next to the pilots' descriptions, we provide an overview of the data that were used as that input for our analysis.

Power Matching City (PMC): Phase 1 (2007–2011) of this pilot [40] consisted of 25 households in the city of Groningen equipped with distributed generators. These generators entail µ-CHP and PV panels. Furthermore, controllable loads were deployed, such as hybrid heat pumps (HP), and smart appliances, such as dishwashers and washing machines. Specific software to match demand and production was developed, i.e., the "PowerMatcher" [45]. In the second phase of the project (2012–2015), more households were incorporated, summing up to a total of 40, and a variable tariff scheme for electricity supply was implemented. For our study, data of 22 households were available. These households had with an average of 3.1 residents per household, with a living area of 150–199 m^2 per household. Typically these numbers are both above the Dutch average [23].

Jouw Energie Moment (JEM): This smart grid pilot project also consisted of two phases. The first phase, 'JEM 1.0', was carried out from 2012 to 2015 [42]. It included 382 residential households in newly constructed buildings in Breda (Easy Street/Meulenspie) and Zwolle (MuziekWijk). All the participants received a smart meter connected to an energy management system (EMS, display in home), indicating dynamic prices, renewable output from rooftop PV, and (for the most part) smart washing machines. In the group of houses in Meulenspie-Breda (39 households), a heat pump was present. A second neighborhood in Breda, in which electricity consumption was monitored, was Easy Street (131 households). As data from Zwolle were not available to us, the number of households was limited to 170 (only Breda). The second phase started in 2017, with higher PV capacity and storage technologies installed. However, this will be the subject of future analyses; it is inappropriate to compare with the two other pilots, as the time period is significantly different [46].

Rendement voor Iedereen (R4I): The Smart Grid: Rendement voor Iedereen (Profit for all) project ran from 2012 to 2014 in two central cities in the Netherlands: Utrecht and Amersfoort. The goal was to develop business cases focused on future smart grids and accompanying energy services. In both cities, smart meters were installed in 100 households that were equipped with solar panels. These meters measured grid power interaction at an unusual high time resolution of 10 s. For the Co-evolution of Smart Energy Products and Services (CESEPS) project, data were collected and analyzed from the Amersfoort pilot [43].

Zonnedael: A smart meter dataset was made available that contained consumption and production values of 70 households in an anonymized neighborhood called Zonnedael in the Netherlands for the year 2013, at a 15 min resolution [44]. It contains a number of PV systems. However, the majority of the households are conventional end users, and no specific energy efficient technologies are included in these households. The dataset is taken as a reference because the energy demand patterns of the households in this dataset resemble those of average Dutch households for that year.

In total, energy data of 287 households have been collected for our analysis.

2.2. Data Properties

In this section the properties of the datasets and their pre-processing are described. Pre-processing was necessary to ensure a certain data quality and monitoring fraction.

Table 1 summarizes the main features and properties of the data. This data consist of time series electrical power, PV production, and household characteristics. For PMC and Zonnedael, the electricity and gas consumption (time series as well as cumulative consumption and production values for PMC) were included. The collected data were registered by smart meters, which were continuously measuring the instantaneous power to average it to energy values over the time resolution (Table 1). The accuracy of all values are the worst case ±2.5%, up to 0.5%, and even lower depending on which meter is chosen, as mentioned in the IEA standards on electrical meters [47]. The differences in the energy systems used in pilots are highlighted. The lowest time resolution was 15 min for JEM, so all other time resolutions are harmonized by using the mean values of 15 min. This leads to a significant effect on the load duration curves and peak analysis but has negligible impact on monthly and annual values.

This analysis is effectuated as a part of CESEPS project [48], and non-disclosure agreements have been signed to access the data (except for Zonnedael). In the EU, the number of starting smart grid pilots and the total investments reached their peaks in 2012, and the number of running projects was the highest in 2013 [11,13]. Therefore, 2013 has been chosen as the year for which data were available for JEM, R4I, and Zonnedael. Nonetheless, for PMC we could only access data for the year 2012. For R4I, data were collected from November 2013 to October 2014. Furthermore, data were not available for all households. From a data availability diagnosis, we took only the data for which the monitoring fraction was higher than 90%. The analysis presented in this paper is based on 21 households for PMC, 79 households for R4I, 117 households for JEM, 26 for Meulenspie, and 91 for Easy Street.

Table 1. Summary of the main features of the smart grid pilots, their data, and the reference dataset.

Residential Smart Meter Data	Location	No.HH *	Time Resolution of Data	PV Systems	HEMS	Energy App	Heat Pumps	m-CHP	Smart Appliances
Jouw energy moment	Breda Easy street	91	15 min		✔	✔			
	Breda Meulenspie	26	15 min	✔	✔	✔	✔		
Powermatching City	Groningen	21	5 min **	✔	✔		✔	✔	✔
Rendement voor Iedereen (Profit for all)	Amersfoort	79	10 s **	✔		✔			
Zonnedael	Anonymized	70	15 min	✔					

* Only households with >90% of the monitoring fraction have been included to the analysis (at 15 min). ** In order to perform the comparison, the time resolution has been reduced to 15 min by using the mean values, to enable the comparison.

2.3. Analysis Indicators and Equations

Each performance indicator will be briefly introduced, and the formulas applied in the analysis will be shown in the section on analysis methods.

For monthly values of energy in kWh, we separately took the sum of the consumption values and production values during the month in question, as shown in Equations (1)–(4):

$$E^i_{C\_m} = \sum_{m\_0}^{m\_end} E^i_C(t) \tag{1}$$

$$E_{C\_m} = \frac{\sum_i^n E^i_{C\_m}}{n} \tag{2}$$

$$E^i_{R\_m} = \sum_{m\_0}^{m\_end} E^i_R(t) \tag{3}$$

$$E_{R\_m} = \frac{\sum_i^n E_{R\_m}^i}{n} \tag{4}$$

where t is the discrete time at 15 min resolution, n is the number of households of the pilot (as cited in Table 1), E_C^i is the consumption of the household i (kWh), $E_{C\_m}^i$ is the total consumption of the household i for the month (kWh), E_{C_m} is the average household consumption value of the pilot for the month in question (kWh), and $m_{\_0}$ and $m_{\_end}$ are the beginning and the end of the month, respectively. At same time step interval, E_R^i is the production of the household i (kWh), $E_{R\_m}^i$ is the total production of the household i, and E_{R_m} is the average household production value of the pilot for the month in question.

For grid operators, besides monthly values, the load duration curve based on residual load, the peak imported electricity, and the PV feed-in and simultanity are also important parameters. These define the capacity that is needed for household connections, cables, and/or transformer substations. The renewable generated electricity subtracted from the electricity consumption provides the hourly residual load, as shown in Equation (5). This gives an indication of the imported and exported electricity to and from the grid and helps to analyze the impacts of new technologies on the grid.

$$P_{residual} = P_C - P_{RE} \tag{5}$$

where $P_{residual}$ is the residual load, P_c is the momentary total electricity consumption in kW, and P_{RE} is the momentary total renewable electricity production in kW. The residual load quantifies the import electricity that the pilot community needs from the grid or the export of renewable generated electricity that the pilot community feeds into the grid. To visualize the residual load throughout the year, a load duration curve is commonly used. This is a curve showing the sorted values of electricity totals, with electricity expressed in power on the y-axis, and its occurrence percentage sorted by power on the x-axis.

Yearly import and export are analyzed according Equations (6) and (7), respectively:

$$\forall t\, E_C^i(t) > E_R^i(t) \;\rightarrow\; E_{imp}^i = \sum_{y\_0}^{y\_end} (E_C^i(t) - E_R^i(t)) \tag{6}$$

$$\forall t\, E_C^i(t) < E_R^i(t) \;\rightarrow\; E_{exp}^i = \sum_{y\_0}^{y\_end} (E_C^i(t) - E_R^i(t)) \tag{7}$$

where $y_{\_0}$ and $y_{\_end}$ are the beginning and the end of the year, E_{imp}^i is the import, and E_{exp}^i is the export of the household i of the pilot in question.

The Self-Consumption Ratio (SCR), as expressed in Equation (8), is the ratio of consumed renewable energy over the sum of all renewable electricity generated on site. Equation (9) formulates the Self-Sufficiency Ratio (SSR), which is the ratio of consumed renewable electricity generated on site over the sum of electricity consumption. The SCR and SSR are calculated according to Equations (8) and (9), respectively:

$$SCR = \frac{E_{C\_RE}}{E_{RE\_tot}} \tag{8}$$

$$SSR = \frac{E_{C\_RE}}{E_{C\_tot}} \tag{9}$$

where E_{C_RE} is the on-site renewable electricity consumption and E_{RE_tot} is the total electricity generated by renewable sources and E_{C_tot} is the total electricity consumption. For this paper, the time period is a year.

The peak load is defined as the maximum electricity import of each household from the grid considering one year of data. Similarly, peak feed-in is the maximum export of each household to the grid. The degree of loads occurring simultaneously is crucial, as individual loads are aggregated on distribution lines and the transformer station. To evaluate the effectiveness of the installed energy systems, it is important to examine the peak load values in smart grid pilots in the Netherlands. In Table 2, we summarize the different residential connections in the Netherlands, regarding their power limits, availability for new connections, and applications.

Table 2. Residential connections and their power limits [49].

Connection	Household Limit (kW)	Appliance Limit (kW) *	Availability	Application
1 × 25 A	5.75	3.68	No longer applied	
1 × 35 A	8.05	7.36	No longer applied	
1 × 40 A	9.2	7.36	Only for small houses	Standard household appliances
3 × 25 A	17.25	11	Standard house connection for new connections	Plus PV panels and EVs
3 × 35 A	24	11	Large connections	Plus PV panels, EVs, and heat pumps

* Considering security plugs of 16 A, which is the standard for the Netherlands [50].

The limit values significantly vary depending on the type of household connection. Exceeding the concerned limit value could have consequences, such as an outage within the household. The connections can always be upgraded for the old and small households if PV or appliances were to be added.

3. Results

In this section, the predefined KPIs analysis' results are presented. The monthly variations underline the difference between technologies. After that, annual energy balance indicators are described, namely, annually imported electricity, the residual load, peak import and export, self-consumption, and self-sufficiency.

3.1. Monthly Electricity Consumption

Figure 1 shows that PV electricity generation is high during summer months for all pilots. For the month of July, the PV generation even matches the total consumption in the JEM-Meulenspie pilot. PV generation is the least for Zonnedael, as only a few households were equipped with PV. Next to the PV system size, the location and the year play a significant role in the electricity produced by PV systems. PMC PV output indicates that the month of May 2012 was relatively sunny. The JEM-Meulenspie PV output shows that, for 2013, the highest solar energy production occurred in July. This result is in line with the weather station data. May 2012 was the month with the highest solar irradiation measured in the weather station that is closest to PMC, i.e., Eelde, with 15.5% of the total irradiation of that year (on a horizontal plane) [51]. An additional effect is the somewhat lower ambient temperatures in May, which resulted in fewer losses compared to higher temperatures. In 2013, July was by far the month with the highest solar irradiation in the weather station closest to Meulenspie, i.e., Gilze-Rijen, with 16.6% of the total irradiation in that year [51].

The consumption patterns are relatively constant for Zonnedael as well as JEM-Easy Street. As mostly students volunteered to participate in the pilot, the lowest consumption occurred at JEM Easy Street. This is clearly related to household size.

Seasonal variations are observed in both JEM-Meulenspie and PMC. In those two pilots, large electricity consumption differences are observed between the winter and summer months due to the heat pumps, which demanded higher amounts of imported electricity during cold periods. The electrical heating resulted in evident variation of consumption patterns. Therefore, for the predefined KPIs, we have analyzed the demand patterns by separating the groups with conventional heating systems from those with hybrid/electric heating systems. Pilots that contain heat pumps are treated separately from those that did not because of their distinct results in power demand.

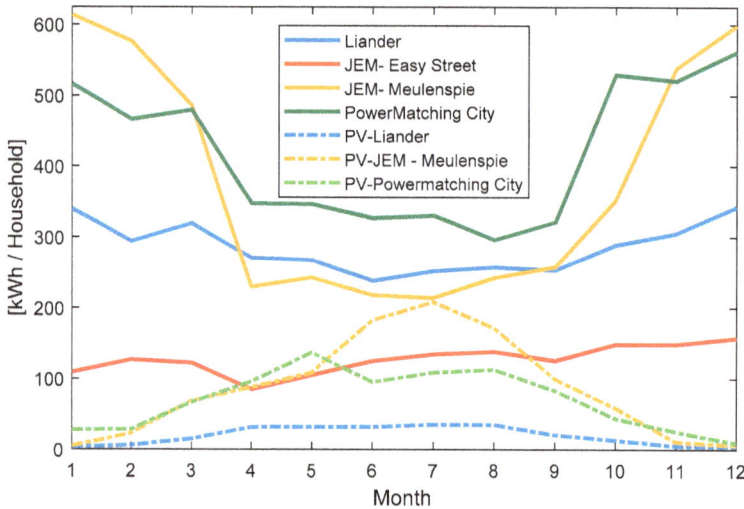

Figure 1. Monthly electricity consumption (solid line) and photovoltaic (PV) production (dashed line) profile over the year for all the datasets. JEM, Jouw Energie Moment.

3.2. Load Duration Curves

Figure 2 shows the aggregated load duration curves, normalized by the number of households, of the smart grid pilots without heat pumps, (i.e., Rendement voor Iedereen and JEM-Easy Street) and the Liander Zonnedael open access dataset t. Figure 3 presents the same for the pilots that included heat pumps (i.e., PMC and JEM-Breda-Meulenspie). For the positive y-axis values, the area between the curve and the x-axis is the electrical energy used per household. The negative y-axis values represent electricity fed into the grid from the PV systems. Only JEM-Easy Street had a lower electricity consumption compared to Zonnedael. From this data, it seems that this pilot effectively achieved its goals of having a low slope of load duration (except for the first 1%), as users were asked to steer on prices, and high prices reflect peak load hours. Nevertheless, the household size is by far the most important parameter to consider: Easy Street in Breda contains apartments inhabited by students, who evidently have a lower energy consumption in general. R4I had a slightly higher consumption for 20% of the time, but, for the remainder of the time, it exhibited a lower consumption than Zonnedael, which can be largely attributed to PV electricity generation in R4I. The peak demand was similar for both data sets.

PMC was characterized by the highest consumption of all smart grid pilots and the highest peak demand, which went up to 5 kW per household (see Figure 3). This is due to the fact that the pilot had a large number of residents per household, a large household area (150–199 m^2) and included heat pumps for space heating. With respect to PV production, R4I had the highest feed-in values of all the pilots, due to the fact that it had the highest share of households with a PV system installed (61 out of the 79 households). This notion is further confirmed by Figure 4, where the boxplots of the yearly household energy demand from the grid and PV power fed in the grid are plotted for all the four datasets.

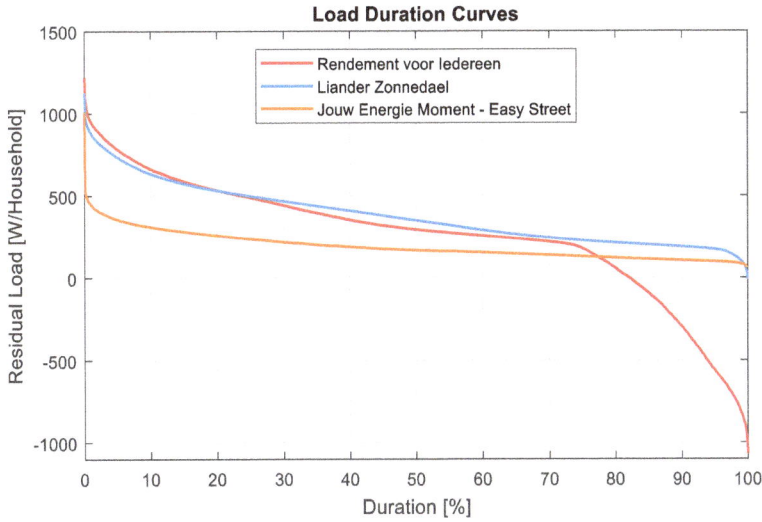

Figure 2. Load duration curves of the "Smartgrid: Rendement voor Iedereen" project, the Liander Zonnedael open access dataset, and the "Jouw Energie Moment" project (exclusively Breda-Easy Street).

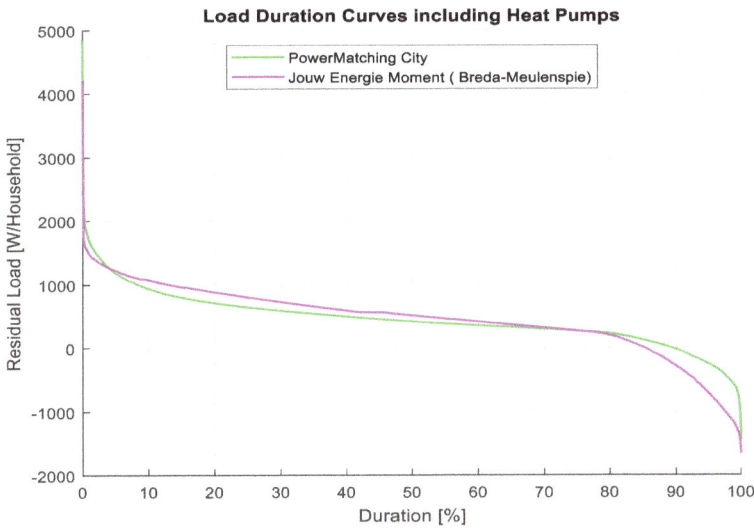

Figure 3. Load duration curves of the "PowerMatching City" project and "Jouw Energie Moment" project—the Meulenspie group (all households are equipped with heat pumps).

3.3. Annualy Imported and Exported Electricity

Figure 4 shows the boxplots of annual imported and exported electricity for all households. For this graph, PMC μ-CHP production is not considered. Most households in R4I have a slightly lower imported electricity than Liander Zonnedael. The average in R4I is 3.01 MWh and in Zonnedael 3.2 MWh. Given that R4I has many PV systems installed, and PV-generated electricity that is directly consumed on-site (i.e., self-consumption), is excluded from the residual load, we can conclude that electricity demand in R4I is probably somewhat above the Dutch average of 3.3 MWh, whereas that of Zonnedael is in line with the average. JEM Easy street is excluded as it has neither heat pumps nor PV.

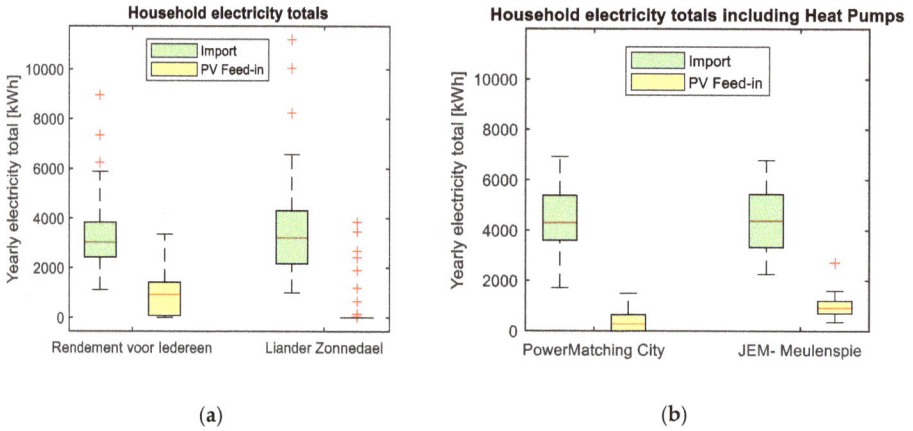

(a)   (b)

Figure 4. Boxplots of annually imported electricity from the grid and surplus PV (feed-in) to the grid for (**a**) household without heat pumps and (**b**) households with heat pumps. For each box, the red line indicates the median, and the bottom and top edges of the box indicate the 25th and 75th percentiles, respectively. The whiskers extend to the most extreme data points without considering the outliers, and the outliers are plotted individually in red.

Figure 4b shows that for the two pilots including heat pumps, JEM, the import is 1 MWh above the Dutch average. These households show a large difference between electricity import and PV feed-in, which means that they are not able to supply their electricity demand with their PV systems.

We observe large differences between the households, both in terms of imported electricity and the amount of PV energy fed into the grid. The upper quartiles show a spread of around 2 MWh. Since most households in the Zonnedael dataset are not equipped with PV systems, all households that do have PV systems are considered outliers for the PV electricity feed-in. Both in imported electricity and the PV energy fed into the grid, we observe large differences between the households, indicated by the maximum and minimum whiskers in the boxplot. This result shows that even households within these small communities are not very similar. This might imply that households with high electricity import may have much potential for energy efficiency measures, whereas households with a low PV feed-in have much potential to increase PV electricity generation. However, a more detailed case-by-case analysis at the household level is required to confirm this.

3.4. Peak Load, Feed-in and Simultaneity

Figure 5 shows the distribution of the power peaks measured for individual households at a time resolution of 15 min and of the normalized peak for the whole community. The latter is the aggregated peak divided by the number of households. JEM-Easy street was not plotted, as few peaks appeared, and PV generation data were absent. Between the pilots analyzed, the peak imported electricity ranged between ~2 kW and ~9 kW for all pilots, with four outliers present above 9 kW.

By looking at an aggregated level, we see an important discrepancy between the households with and without heat pumps. The average peak per household is only around 1 kW for pilots without heat pumps, which is in line with how transformers are generally dimensioned [52]. If we define the simultaneity of electricity as the ratio between the normalized aggregated peak and the average individual peaks, then we observe a very low import simultaneity in the R4I pilot of only 23.0%. This is beneficial from a system perspective, as this low import simultaneity decreases the total capacity requirement. The same holds for Zonnedael, with a simultaneity of 20.5%. However, the simultaneity in the pilots with heat pumps is much higher: 65.1% for JEM and 70.5% for PMC. This can be explained by the fact that heat pumps' electricity demands are strongly correlated with outside temperatures, which are the same for all households within a community. While, on average, the electricity demand of households with heat pumps is just around 200 Watts higher than that of households without heat pumps, the community peak exceeds 3 kW on top of the baseline. This causes a stress in the LV grid and its transformers. Therefore, it is important to take this into account when designing future LV grids with a high penetration of heat pumps.

Not surprisingly, the simultaneity of the PV feed-in is also very high: 74.9% for the R4I pilot and 89.5% for JEM (very similar to Zonnedael 88.9%). This calculation was not applicable for PMC, where the virtual PV was controlled by the DSO and distributed to households with PowerMatcher influence, since we noticed that not all households received the same amount of electricity. It is important to mention that, in the investigated pilots, these results show a low potential to cause grid issues: all aggregated feed-in peaks are lower than the aggregated import peaks. Furthermore, feed-in peaks are not as crucial as import peaks, as it is easier to curtail feed-in [32].

From the investigated connections, only four households may present a possible risk for overloading (i.e., those with a 1×40 A connection). We do note, however, that this connection is applied only for small houses on special request. For the standard house connection, there were no households who exceeded those levels.

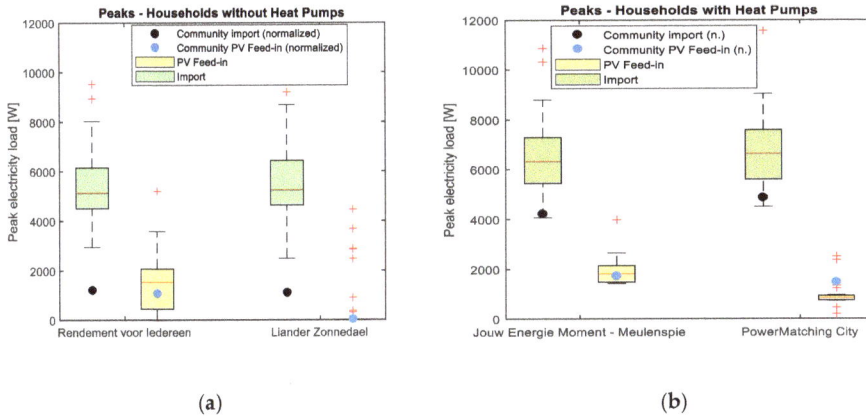

(a)   (b)

Figure 5. Boxplots of annually imported electricity from the grid and surplus of PV generated electricity (feed-in) to the grid, based on 15 min averages. (**a**) results for Rendement voor Iedereen pilot and Liander (without electrical heating), (**b**) results for Jouw Energie Moment-Meulenspie and Power Matching City (with electrical heating). For each box, the red line indicates the median, and the bottom and top edges of the box indicate the 25th and 75th percentiles, respectively. The whiskers extend to the most extreme data points without considering the outliers, and the outliers are plotted individually in red. The black dots represent the normalized import peak for the whole community, and the blue dot represents the normalized PV feed-in peak.

3.5. Self-Consumption and Self-Sufficiency

With our limited amount of data, we could only calculate self-consumption and self-sufficiency ratios for R4I and PMC for 2012, which is shown in Figure 6 for all households. Again, we observe a large difference between the households, with a median SCR of around 50% for the R4I pilot and 70% for PMC. The general picture shows relatively high SCRs. This can be explained by the relatively small PV system sizes in these pilots. With fewer PV electricity generated, larger parts of the generated electricity can be directly consumed. Another notion is that the aggregated values are much higher (5–20%). This means that substantial parts of the PV electricity fed-in by households are used within the community. This is of importance, as this electricity is not exported to the MV-grid, thereby decreasing the overall stress on the electricity system.

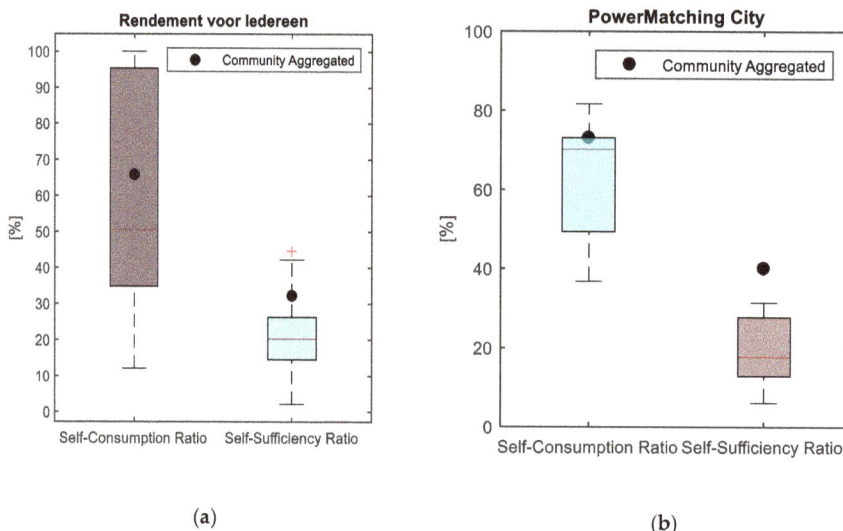

(a)   (b)

Figure 6. Boxplots of annual self-consumption and self-sufficiency ratios. (**a**) results for Rendement voor Iedereen pilot and (**b**) for Power Matchig City. For each box, the red line indicates the median, and the bottom and top edges of the box indicate the 25th and 75th percentiles, respectively. The whiskers extend to the most extreme data points without considering the outliers, and the outliers are plotted individually in red. The black dots represent the community aggregated values for self-consumption and self-sufficiency.

3.6. Summary of Results

Table 3 summarizes the main results presented in the previous sections. Much of the data was available for only a small number of households, leading to low statistical power. Therefore, it is difficult to draw generic conclusions, especially considering variable household sizes, differences in the technologies deployed, and other socio-economic factors. Nevertheless, the boxplots results give insight into the various scenarios that occurred within the smart grid pilots, and the average results provide an insight into the technologies' impact.

Table 3. Summary of the calculated key performance indicators in average values.

Smart Meter Data	Annual Import (MWh)	Peak Import (kW)	Peak PV (kW)	Annual PV Generation (MWh)	Self-Consumption (%)	Self-Sufficiency (%)	Import Simultaneity (%)	Feed-in Simultaneity (%)
Zonnedael	3.2	4.8	0.27	0.22	-	-	20.5%	88.8%
R4I	3.1	5.1	2.2	1	50%	20%	23.0%	74.9%
JEM-Meulenspie	4.5	6.2	1.8	0.9	-	-	65.1%	89.5%
PMC	4.3	6.8	0.5	0.86	70%	18%	70.5%	n.a.

4. Discussion

From our results, it can be concluded that simultaneity in electricity generation or load is much higher for energy technologies that are being deployed more and more frequently in the context of energy transition, namely PV systems [53] and heat pumps. This should be an important focus point for DSOs. The power flows found in this study could be hazardous if PV systems and heat pumps were to be installed in all Dutch households, as current grids in the Netherlands are not equipped to transport the reported power flows. Various options exist to address this challenge. For PV, one option is to curtail electricity generation during hours with high solar irradiation. This only leads to limited losses in the total energy yield [32]. A second option is to store the surplus of PV-generated electricity in storage, where stationary storage and EVs are options. In the latter case, smart solar charging holds much potential. Heat pumps present an even larger challenge, as controlling these loads could have an impact on the comfort levels of households. One option is to provide incentives to households to increase the range of temperatures that households would tolerate. Also, electricity could be supplied by stationary or electric vehicle storage. Lastly, thermal storage could be an attractive option. Future analyses must determine which option, or which combination of options, is most suitable. Further, we will quantitatively discuss the main findings by categories.

Regarding the consumption patterns of our sample, the yearly exported electricity values are quite modest, considering the fact that not everyone in a neighborhood owns a PV system, Furthermore, results show that most PV generated electricity is consumed locally without resulting into transmission losses. This holds true even for communities with a high penetration of PV systems. R4I had an average PV system size per household of 1.7 kWp, but SSR at the community level was still 60–70%. Yearly imported values have a diverse range considering the size of the samples. For creating insight into various consumption patterns, pilots usually try to include various household sizes to sample different fractions of the population. For instance, this is not the case in JEM-Easy street, which ends up having outstandingly low consumption, which is not representative for an average household in the Netherlands. However, other groups or pilots had more diverse household sizes. For example, JEM-Meulenspie and PMC included heat pumps, which made the annual electricity consumption vary between 2 to 8 MWh, while the electricity consumption without heat pumps (in R4I) varied between 1 to 6 MWh in their annual imported electricity (Figure 4). Caution should be taken while scaling up these results. The socio-economical properties, household sizes and surfaces, types of households, technologies employed, and geographical attributes should be similar within a region, to get as representative a result as possible.

Considering the load duration curves and peak PV feed-in values, an increased delivery of electricity to the grid can also have similar consequences for power grid failures. With respect to the solar PV fed into the grid, R4I showed the largest values (Figure 5). The second largest delivery of renewable power to the grid was realized in JEM-Meulenspie. PMC had a virtual PV system for 18 households, with power delivered, controlled, and distributed by an energy algorithm, which made the distribution narrower. The physical PV rooftop installations are the outliers. That is most probably why the community aggregated normalized in feed peaks is more than the average at that particular moment. In Zonnedael there are few households with PV systems, but these are considered outliers.

Concerning outliers, for the households without heat pumps, even the highest outliers are safe in the sense that they do not pose a risk for outages given their relatively low values. Within the sample of 47 households with heat pumps, only three households show a peak over 10 kW. These households most likely have a large connection, whereas the new standard connection allows their safe operation even in extreme scenarios. In the case of the use of heat pumps, it is recommended to integrate an energy management system for managing activation time, taking into account neighborhood consumption. As the electricity demand of heat pumps is highly dependent on the ambient temperature, these systems could exhibit a high simultaneity in demand, and if not well managed, this simultaneity might result in peak demand at the same time, and in a high load of transformers in some extreme scenarios, even though such circumstances did not occur during the investigated time period.

5. Conclusions

In this paper, we have analyzed energy data from 287 households from five different locations to determine seven key performance indicators: monthly electricity consumption (range of pilots averages on household level: 100–600 kWh) and production (idem: 4–200 kWh), annually imported (idem: 3.1–4.5 MWh) and exported (idem: 0.2–1 MWh) electricity, residual load, peak of imported (idem: 4.8–6.8 kW) and exported (idem: 0.3–2.2 kW) electricity, import simultaneity (idem: 20–70.5%) and feed-in simultaneity (idem: 75–89%), self-sufficiency (idem: 18–20%), and (7) self-consumption (idem: 50–70%).

Our results not only identified the average household of the pilots, diagnosed the extrema as the households with outlier characteristics and peak loads, but also distinguished the community values. On community level, SCRs and SSRs are noticeably higher, showing that households are able to supply part of the electricity need of their neighbors. Local use of PV-generated electricity decreases the stress on the LV/MV transformer and accompanying transmission losses. Apart from the load curves shown in this paper, it was found that the electrification of heating systems in buildings which use heat pumps, leads to an increase of their annual electricity consumption and peak loads of approximately 30%, compared to average Dutch households without heat pumps. Further it the investigated households showed a low self-sufficiency of around 20%, mainly because of their use of relatively small PV systems. Therefore, they could not be classified as energy neutral. To increase the self-sufficiency of households, further investigations are required to optimize smart grid systems. It becomes apparent that households shifting to electric heating (heat pumps) will face a significant increase in electricity consumption compared to the Dutch average. This is quite relevant for the Netherlands, as all newly built buildings are not allowed to be connected to the gas network anymore due to the Climate Agreement (klimaatakkoord). The integration of heat pumps will also have implications for network operators in terms of increased costs for reinforcing the grid to accommodate increasing demand. Especially attributed to the high simultaneity of consumption of heat pumps due to correlation with ambient temperature. Relatively low self-sufficiency ratios and high self-consumption ratios have been obtained in the pilots, since the installed PV capacity was rather small. Self-sufficiency is expected to increase with the current trend of installing larger solar installations on roof tops. In perspective, integration and size of the battery will be part of a future work, by using better monitoring and more households' data to increase statistical significance.

Author Contributions: Writing–original draft preparation C.G.; formal analysis, investigation, and algorithms, C.G. and W.S.; conceptualization and validation, C.G., W.S., A.R., I.L., and W.v.S.; resources, review, editing, A.R., I.L., and W.v.S.; supervision and data acquisition, A.R. and W.v.S.

Funding: This research has received funding from the European Union's Horizon 2020 research and innovation programme under the ERA-Net Smart Grids plus, grant number 646039, from the Netherlands Organisation for Scientific Research (NWO).

Acknowledgments: We would like to acknowledge all participants in the smart grid pilots (PowerMatching City, Jouw Energie Moment, Rendement voor Iedereen) involved in this study for their willingness to share their data, experiences, and knowledge with the researchers. We would like to thank Daphne Geelen, Frits Bliek and Alexander Savelkouls for facilitating the data acquisition, and last but not least, our colleagues Carla Robledo, Ad van Wijk and Barbara van Mierlo for their kind advices and suggestions.

Conflicts of Interest: The authors declare no conflict of interest.

Disclaimer: The content and views expressed in this material are those of the authors and do not necessarily reflect the views or opinion of the ERA-Net SG+ initiative. Any reference given does not necessarily imply the endorsement by ERA-Net SG+.

References

1. EU Commission. *Smart Grids: From Innovation to Deployment*; European Commission: Brussels, Belgium, 2011; pp. 1–13.

2. Zunnurain, I.; Maruf, M.N.I.; Rahman, M.M.; Shafiullah, G. Implementation of Advanced Demand Side Management for Microgrid Incorporating Demand Response and Home Energy Management System. *Infrastructures* **2018**, *3*, 50. [CrossRef]

3. Lampropoulos, I.; Alskaif, T.; van den Broek, M.; van Sark, W.; van Oostendorp, H. A Method for Developing a Game-Enhanced Tool Targeting Consumer Engagement in Demand Response Mechanisms. In *Mediterranean Cities and Island Communities*; Stratigea, A., Kavroudakis, D., Eds.; Springer International Publishing: Cham, Germany, 2019; pp. 213–235. ISBN 978-3-319-99443-7.

4. Posma, J.; Lampropoulos, I.; Schram, W.; Van Sark, W. Provision of Ancillary Services from an Aggregated Portfolio of Residential Heat Pumps on the Dutch Frequency Containment Reserve Market. *Appl. Sci.* **2019**, *9*, 590. [CrossRef]

5. Lampropoulos, I.; AlSkaif, T.; Blom, J.; Van Sark, W. A framework for the provision of flexibility services at the transmission and distribution levels through aggregator companies. *Sustain. Energy Grids Netw.* **2019**, *17*, 100187. [CrossRef]

6. Lampropoulos, I.; Kling, W.L.; Garoufalis, P.; Bosch, P.P.V.D. Hierarchical predictive control scheme for distributed energy storage integrated with residential demand and photovoltaic generation. *IET Gener. Transm. Distrib.* **2015**, *9*, 2319–2327. [CrossRef]

7. Lampropoulos, I.; Veldman, E.; Kling, W.L.; Gibescu, M.; Slootweg, J.G. Electric vehicles integration within low voltage electricity networks & possibilities for distribution energy loss reduction. In Proceedings of the Proc. 2010 Innovation for Sustainable Production (i-SUP), Sustainable Energy Conf. (3), Brudges, Belgium, 18–21 April 2010; pp. 74–78.

8. Lampropoulos, I.; Broek, M.V.D.; Van Der Hoofd, E.; Hommes, K.; Van Sark, W. A system perspective to the deployment of flexibility through aggregator companies in the Netherlands. *Energy Policy* **2018**, *118*, 534–551. [CrossRef]

9. Lampropoulos, I.; van den Broek, M.; van Sark, W.; van der Hoofd, E.; Hommes, K. Enabling Flexibility from Demand-Side Resources Through Aggregator Companies. In *Smart Cities in the Mediterranean*; Stratigea, A., Kyriakides, E., Nicolaides, C., Eds.; Springer International Publishing: Cham, Switzerland, 2017; pp. 333–353. ISBN 978-3-319-54557-8.

10. Lampropoulos, I.; Baghina, N.; Kling, W.L.; Ribeiro, P.F. A Predictive Control Scheme for Real-Time Demand Response Applications. *IEEE Trans. Smart Grid* **2013**, *4*, 2049–2060. [CrossRef]

11. Covrig, C.F.; Ardelean, M.; Vasiljevska, J.; Mengolini, A.; Fulli, G.; Amoiralis, E.; Jiménez, M.S.; Filiou, C. *Smart Grid Projects Outlook 2014*; Joint Research Centre of the European Commission: Petten, The Netherlands, 2014.

12. Joint Research Center of European Commission Smart Grid Project List. Available online: https://ses.jrc.ec.europa.eu/inventory?field_proj_dev_stage_value=All&field_proj_start_date_value%5Bvalue%5D%5Byear%5D=&field_proj_start_date_value2%5Bvalue%5D%5Byear%5D=&field_proj_countries_involed_tid=&titleproj=&field_proj_application_value= (accessed on 29 March 2019).

13. JRC. *Smart Grid Project List*; European Commission: Brussels, Belgium, 2019.

14. Ministry of Economic Affairs. *Position paper kennis- en leertraject Thema Wet- en Regelgeving*; Rijksdienst voor Ondernemend Nederland: Utrecht, The Netherlands, 2015.

15. Rijksoverheid. *Gouverment of the Netherlands Experimentation Decree for Experiments with Decentralized Sustainable Electricity Generation (Besluit Experimenten Decentrale Duurzame Elektriciteitsopwekking)*; Article 1–18; Rijksoverheid: The Hague, The Netherlands, 2015.

16. Lammers, I.; Diestelmeier, L. Experimenting with Law and Governance for Decentralized Electricity Systems: Adjusting Regulation to Reality? *Sustainability* **2017**, *9*, 212. [CrossRef]

17. Ministry of Economic Affairs and Climate Policy. *Dutch Goals with EU*; Ministry of Economic Affairs and Climate Policy: The Hague, The Netherlands, 2017.

18. Climate Council Klimaatakkoord. Available online: https://www.klimaatakkoord.nl/ (accessed on 29 March 2019).

19. Geelen, D.V. *Empowering End-Users in the Energy Transition: An Exploration of Products and Services to Support Changes in Household Energy Management*; TU Delft: Delft, The Netherlands, 2014.

20. Hansen, M.; Borup, M. Smart grids and households: How are household consumers represented in experimental projects? *Technol. Anal. Strategic Manag.* **2018**, *30*, 255–267. [CrossRef]

21. Geelen, D.; Scheepens, A.; Kobus, C.; Obinna, U.; Mugge, R.; Schoormans, J.; Reinders, A. Smart energy households' pilot projects in The Netherlands with a design-driven approach. In Proceedings of the IEEE PES ISGT Europe 2013, Lyngby, Denmark, 6–9 October 2013; IEEE: Lyngby, Denmark, 2013; pp. 1–5.

22. Schram, W.L.; Lampropoulos, I.; van Sark, W.G. Photovoltaic systems coupled with batteries that are optimally sized for household self-consumption: Assessment of peak shaving potential. *Appl. Energy* **2018**, *223*, 69–81. [CrossRef]

23. Gercek, C.; Reinders, A. Smart Appliances for Efficient Integration of Solar Energy: A Dutch Case Study of a Residential Smart Grid Pilot. *Appl. Sci.* **2019**, *9*, 581. [CrossRef]

24. Brouwers, H.; van Mierlo, B. *Residential Smart Grid Projects in the Netherlands: An Overview*; Wageningen University & Research: Wageningen, The Netherlands, 2019.

25. Zhou, K.; Yang, S.; Shen, C. A review of electric load classification in smart grid environment. *Renew. Sustain. Energy Rev.* **2013**, *24*, 103–110. [CrossRef]

26. Pramangioulis, D.; Atsonios, K.; Nikolopoulos, N.; Rakopoulos, D.; Grammelis, P.; Kakaras, E. A Methodology for Determination and Definition of Key Performance Indicators for Smart Grids Development in Island Energy Systems. *Energies* **2019**, *12*, 242. [CrossRef]

27. Lampropoulos, I.; Vanalme, G.M.A.; Kling, W.L. A methodology for modeling the behavior of electricity prosumers within the smart grid. In Proceedings of the 2010 IEEE PES Innovative Smart Grid Technologies Conference Europe (ISGT Europe), Gothenburg, Sweden, 11–13 October 2010; IEEE: Gothenburg, Sweden, 2010; pp. 1–8.

28. Brouwers, H.; Gültekin, E.; Mierlo, B. *Learning about User Engagement in Smart Grid Niche Development: An Analysis of 4 Smart Grid Projects*; Wageningen University & Research: Wageningen, The Netherlands, 2019.

29. Personal, E.; Guerrero, J.I.; Garcia, A.; Peña, M.; Leon, C. Key performance indicators: A useful tool to assess Smart Grid goals. *Energy* **2014**, *76*, 976–988. [CrossRef]

30. Litjens, G.B.M.A.; Worrell, E.; van Sark, W.G.J.H.M. Lowering greenhouse gas emissions in the built environment by combining ground source heat pumps, photovoltaics and battery storage. *Energy Build.* **2018**, *180*, 51–71. [CrossRef]

31. Bossmann, T.; Barberi, P.; Fournié, L. *Effect of High Shares of Renewables on Power Systems*; EU Commission: Brussel, Belgium, 2018; pp. 1–34.

32. Litjens, G. *Here Comes the Sun: Improving Local Use of Electricity Generated by Rooftop Photovoltaic Systems*; Utrecht University: Utrecht, The Netherlands, 2018.

33. Litjens, G.B.M.A.; Worrell, E.; van Sark, W.G.J.H.M. Assessment of forecasting methods on performance of photovoltaic-battery systems. *Appl. Energy* **2018**, *221*, 358–373. [CrossRef]

34. Litjens, G.B.M.A.; Worrell, E.; van Sark, W.G.J.H.M. Economic benefits of combining self-consumption enhancement with frequency restoration reserves provision by photovoltaic-battery systems. *Appl. Energy* **2018**, *223*, 172–187. [CrossRef]

35. Luthander, R.; Widén, J.; Nilsson, D.; Palm, J. Photovoltaic self-consumption in buildings: A review. *Appl. Energy* **2015**, *142*, 80–94. [CrossRef]

36. Van der Stelt, S.; AlSkaif, T.; van Sark, W. Techno-economic analysis of household and community energy storage for residential prosumers with smart appliances. *Appl. Energy* **2018**, *209*, 266–276. [CrossRef]

37. Cook, J.; Ardani, K.; O'Shaughnessy, E.; Smith, B.; Margolis, R. *Expanding PV Value: Lessons Learned from Utility-led Distributed Energy Resource Aggregation in the United States*; National Renewable Energy Laboratory: Golden, CO, USA, 2018; pp. 1–37.

38. Shoeb, M.A.; Shahnia, F.; Shafiullah, G. A Multilayer and Event-triggered Voltage and Frequency Management Technique for Microgrid's Central Controller Considering Operational and Sustainability Aspects. *IEEE Trans. Smart Grid* **2018**, 1. [CrossRef]

39. Bliek, F.; van den Noort, A.; Roossien, B.; Kamphuis, R.; de Wit, J.; van der Velde, J.; Eijgelaar, M. PowerMatching City, a living lab smart grid demonstration. In Proceedings of the 2010 IEEE PES Innovative Smart Grid Technologies Conference Europe (ISGT Europe), Gothenburg, Sweden, 11–13 October 2010; IEEE: Gothenburg, Sweden, 2010; pp. 1–8.

40. PowerMatching City A Demonstration Project of a Future Energy Infrastructure. Available online: http://powermatchingcity.nl/ (accessed on 12 April 2018).

41. Geelen, D.; Vos-Vlamings, M.; Filippidou, F.; van den Noort, A.; van Grootel, M.; Moll, H.; Reinders, A.; Keyson, D. *An End-User Perspective on Smart Home Energy Systems in the PowerMatching City Demonstration Project*; IEEE: Lyngby, Denmark, 2013; pp. 1–5.

42. Jouw Energie Moment Keeping Energy Affordable and Available. Available online: http://www.jouwenergiemoment.nl/jouw-energie-moment/ (accessed on 26 November 2018).

43. Economic Board Utrecht (EBU) Smart Grid: Rendement Voor Iedereen. Available online: https://energiekaart.net/initiatieven/smart-grid-rendement-voor-iedereen/ (accessed on 26 November 2018).

44. Liander Smart Meter Data. Available online: https://www.liander.nl/partners/datadiensten/open-data/data (accessed on 29 March 2019).

45. Kok, J.K.; Scheepers, M.J.J.; Kamphuis, I.G. Intelligence in Electricity Networks for Embedding Renewables and Distributed Generation. In *Intelligent Infrastructures*; Negenborn, R.R., Lukszo, Z., Hellendoorn, H., Eds.; Springer Netherlands: Dordrecht, The Netherlands, 2010; pp. 179–209, ISBN 978-90-481-3597-4.

46. Jouw Energie Moment. *Jouw Energie Moment—Geaggregeerde Opslag*; Enexis, Technolution, Senfal: Breda, The Netherlands, 2018; p. 37.

47. International Electrotechnical Commission 62053-11. *Electricity Metering Equipment (A.C.)—Particular Requirements—Part 11: Electromechanical Meters for Active Energy (Classes 0.5, 1 and 2)*; IEC: Geneva, Switzerland, 2003.

48. Reinders, A.; Übermasser, S.; van Sark, W.; Gercek, C.; Schram, W.; Obinna, U.; Lehfuss, F.; van Mierlo, B.; Robledo, C.; van Wijk, A. An Exploration of the Three-Layer Model Including Stakeholders, Markets and Technologies for Assessments of Residential Smart Grids. *Appl. Sci.* **2018**, *8*, 2363. [CrossRef]

49. Liander Types of Connections for Gas and Electricity. Available online: https://www.liander.nl/consument/aansluitingen/typen (accessed on 24 January 2018).

50. IEC TR 60083. *Plugs and Socket-Outlets for Domestic and Similar General Use Standardized in Member Countries of IEC*; IEC: Geneva, Switzerland, 2015.

51. Royal Netherlands Meteorological Institute (KNMI)) Hourly Weather Data in the Netherlands. Available online: https://projects.knmi.nl/klimatologie/uurgegevens/selectie.cgi (accessed on 2 April 2019).

52. Van Oirsouw, P.; Cobben, J.F.G. *Netten Voor Distributie van Elektriciteit: Phase to Phase*; Alliander and Agentschap NL: Arhnem, The Netherlands, 2011.

53. Reinders, A.; Verlinden, P.; Van Sark, W.; Freundlich, A. *Photovoltaic Solar Energy: From Fundamentals to Applications*; John Wiley & Sons: Hoboken, NJ, USA, 2017; ISBN 978-1-118-92747-2.

![applied sciences logo] *applied sciences*

MDPI

Article

Simulation-Supported Testing of Smart Energy Product Prototypes

Alonzo Sierra [1,*], Cihan Gercek [1], Stefan Übermasser [2] and Angèle Reinders [1,3]

[1] Department of Design, Production and Management, Faculty of Engineering Technology, University of Twente, P.O. Box 217, 7500 AE Enschede, The Netherlands; c.gercek@utwente.nl (C.G.), a.h.m.e.reinders@utwente.nl (A.R.)

[2] Austrian Institute of Technology, Giefinggasse 4, 1210 Vienna, Austria; stefan.uebermasser@ait.ac.at

[3] Energy Technology Group at Mechanical Engineering, Eindhoven University of Technology, P.O. Box 513, 5600 MB Eindhoven, The Netherlands

* Correspondence: a.sierrarodriguez@utwente.nl; Tel.: +31-617134264

Received: 9 April 2019; Accepted: 15 May 2019; Published: 17 May 2019

Abstract: Smart energy products and services (SEPS) have a key role in the development of smart grids, and testing methods such as co-simulation and scenario-based simulations can be useful tools for evaluating the potential of new SEPS concepts during their early development stages. Three innovative conceptual designs for home energy management products (HEMPs)—a specific category of SEPS—were successfully tested using a simulation environment, validating their operation using simulated production and load profiles. For comparison with reality, end user tests were carried out on two of the HEMP concepts and showed mixed results for achieving more efficient energy use, with one of the concepts reducing energy consumption by 27% and the other increasing it by 25%. The scenario-based simulations provided additional insights on the performance of these products, matching some of the general trends observed during end user tests but failing to sufficiently approximate the observed results. Overall, the presented testing methods successfully evaluated the performance of HEMPs under various use conditions and identified bottlenecks, which could be improved in future designs. It is recommended that in addition to HEMPs, these tests are repeated with different SEPS and energy systems to enhance the robustness of the methods.

Keywords: smart product design; smart home technology; power systems simulation; energy management

1. Introduction

Within the framework for the development and implementation of smart grids, smart energy products and services (SEPS) are set to play a key role. SEPS are solutions "expected to support the active participation of end users in balancing energy demand and supply in the electricity network" [1] by creating an environment where energy use is flexible [2–4], efficient, reliable [5], sustainable and cost-effective [6]. Examples of SEPS include smart meters, smart appliances, electric and fuel cell vehicles [7,8], residential energy storage systems [9,10], and home energy management systems (HEMS) [11] among others.

The widespread implementation of SEPS in smart grids could enable greater interaction between end users, home appliances and energy suppliers, facilitating energy efficiency, local production and energy trading with the grid in order to improve the effectiveness of demand response strategies and reduce the required capacity for local energy storage [12,13]. This requires more active end user involvement which is currently limited by user acceptance, with users frequently finding SEPS difficult to understand and interact with [14–16]. Therefore, there is a need to develop more innovative SEPS that facilitate this role by achieving a better match with user expectations and demands.

Testing methods such as co-simulation and scenario-based simulations can be useful tools for quickly evaluating the technical functioning and preferred user interaction with a new SEPS design during its early development stages. Co-simulation is a method where several subsystems are simulated independently and then coupled together to analyse the entire system and the interactions between its components. This makes it possible to quickly and accurately model complex, heterogeneous systems by using the simulating tools native to each subsystem [17,18]. The use of co-simulations for modelling smart energy systems has been explored in the literature [18–20] but there is still little evidence of its application with SEPS prototypes. Scenario-based simulations, on the other hand, can use existing energy profile datasets to replicate real-life conditions without the need to carry out field tests, although it is important that the used data accurately reflects system behaviour observed in practice. Relevant examples of these tests applied to smart energy systems include the scenario-based simulations found in [21–23] as well as the user tests presented by [24,25].

In this paper, the performance of three home energy management product (HEMP) concepts are tested using a simulation testing environment as well as scenario-based simulations; results from end user tests are presented as well, serving as a validation for the modelled scenarios. The article is structured as follows: Section 2 introduces the methodology followed for each of the proposed testing methods, and the results of each test are presented in Section 3. In Section 4, a discussion on the effectiveness of these methods is presented followed by some conclusions on this study.

2. Materials and Methods

Prototypes for three conceptual HEMP designs were specifically developed for this study to serve as user interfaces for home energy management. They consisted of devices with visually appealing forms that measure energy production and consumption data from a household smart meter and display basic information to users through simple, intuitive visual feedback such as LED colouring and brightness. This feedback is updated on regular intervals to indicate how household performance changes through time and in response to users' actions. The three developed concepts (shown in Figure 1) are described below:

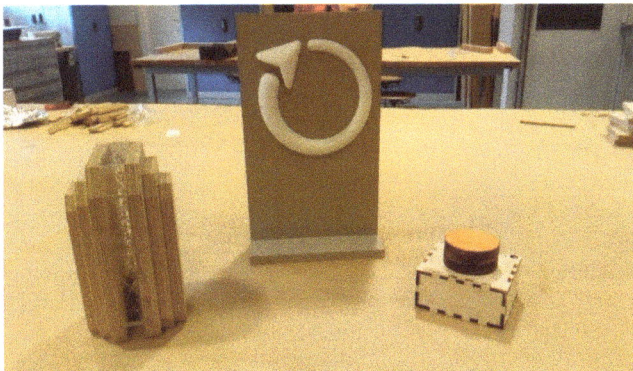

Figure 1. The three developed SEPS prototypes: CrystalLight (left), Bodhi (centre) and LightInsight (right).

1. **Bodhi:** An arrow-shaped "energy budget" indicator that shows users how a household's energy use compares to a predetermined daily or weekly budget through LED colouring. The relationship between actual and planned consumption during a given interval is determined through a budget ratio (R_B, unitless) defined as:

$$R_B = E_{cum}/(j/N) * B, \tag{1}$$

where E_{cum} (kWh) is the cumulative energy consumption in the current period (e.g., a day or a week), j is the interval number (unitless), N is the number of intervals in a period (unitless), and B is the total energy budget for a given period (kWh). An R_B value between 0.95 and 1.05 indicates that users are "on budget" (corresponding to purple LED lighting); values greater than 1.05 correspond to an "over budget" state (orange lighting) while values below 0.95 indicate the household is "under budget" (aqua lighting).

2. **CrystalLight**: A smart home ornament that acts like a virtual energy storage system; each day, electricity produced by a household's PV array makes its LEDs grow stronger ("charging" the ornament) while electricity consumption gradually dims them. The charge (C_i, kWh) and state of charge (SOC_i, unitless) for this "battery" at each measurement interval are calculated as:

$$C_i = C_{i-1} + E_P - E_C \tag{2}$$

and:

$$SOC_i = C_i/C_{MAX}, \tag{3}$$

where E_P (kWh) and E_C (kWh) are the produced and consumed energy during a given interval and C_{MAX} denotes the total battery capacity (kWh). The battery's state of charge is converted into a brightness value between 0 and 100% for the prototype lights; if at any given interval C_i becomes negative, it will be automatically set to zero to simulate an "empty" battery. Likewise, if the state of charge becomes greater than 100%, the charge will be set to C_{MAX} to simulate a "full" battery.

3. **LightInsight**: A small cylindrical dial that gives users information on the balance between a household's energy production and consumption during the day through LED lighting. Four different feedback states are possible: net energy production (green lighting), net consumption (red), transition from green to red (yellow) and transition from red to green (rainbow). These states are determined by an energy ratio (R_E, unitless) defined as:

$$R_E = E_P/E_C, \tag{4}$$

where E_P (kWh) and E_C (kWh) are the produced and consumed energy during a given interval.

The autonomous operation of all prototypes was made possible through the use of Raspberry Pi microprocessors, where a Python script was created to periodically obtain energy data from a smart meter, calculate the required key indicator (R_B, SOC_i or R_E) and set the LED properties accordingly.

2.1. Simulation Environment Testing

A series of short testing sequences were developed to validate prototype operation using the simulation environment from the Smart Electricity Systems and Technology Services laboratory (SmartEST Lab) at the Austrian Institute of Technology (AIT) [26]. In these tests, energy production and consumption were independently simulated to model different system states, which were interpreted by each prototype in order to set its LED properties accordingly, as seen in Figure 2 below. This was achieved through the following process:

- Energy **generation** was modelled using a DC voltage/current source, which simulated a residential PV system. Energy **consumption**, on the other hand, was modelled using an RLC controllable load, which consumed the generated power or drew power from the local grid whenever consumption exceeded generation.
- The laboratory's main measurement system integrated these two inputs and periodically passed them on to the HEMP prototype using the communication infrastructure, which consisted of a custom-built middleware application linking these components.
- The prototype calculated the key indicator's new value and set the corresponding LED properties.

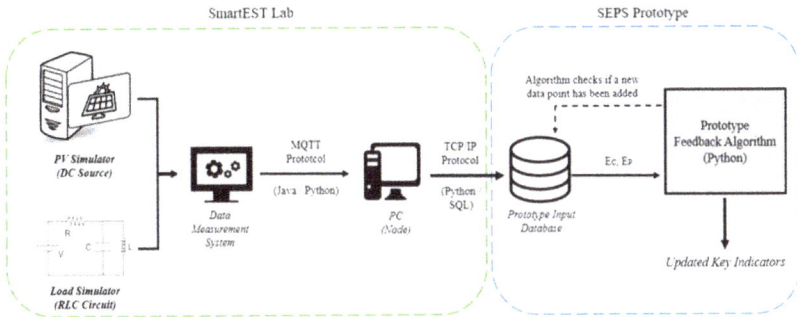

Figure 2. Test setup for the HEMP test sequences in the AIT simulation environment.

2.2. Scenario-Based Simulations

The operation of each prototype was further tested by using existing production and consumption datasets to model several use scenarios. Four different scenarios were created by combining summer and winter load curves with PV production data reflecting "adequate" or "inadequate" performance according to weather conditions; all sources have 1-min resolution and cover a 24-h period as seen in Figure 3 below. The following scenarios were modelled:

1. Summer Load Profile, Inadequate PV Production
2. Summer Load Profile, Adequate PV Production
3. Winter Load Profile, Inadequate PV Production
4. Winter Load Profile, Adequate PV Production

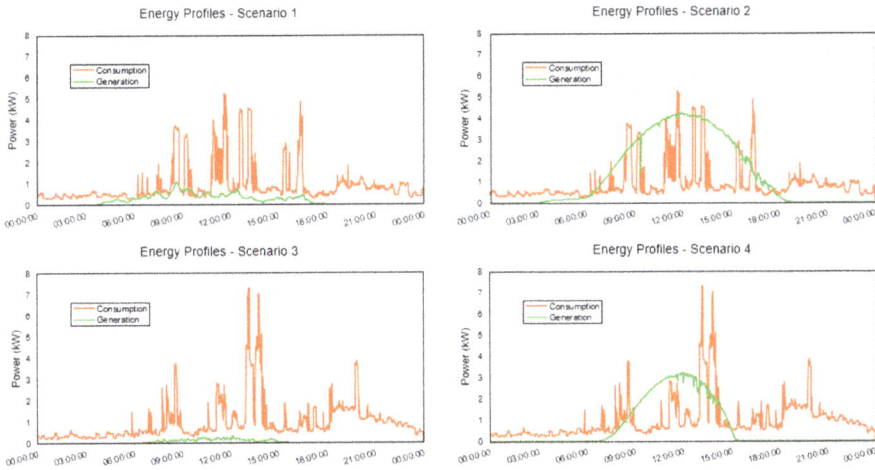

Figure 3. Energy profiles for the four modelled scenarios, showing consumption in orange and production in green. Clockwise from top left: Scenario 1 (Summer Load, Inadequate PV), Scenario 2 (Summer Load, Adequate PV), Scenario 4 (Winter Load, Adequate PV), Scenario 3 (Winter Load, Inadequate PV).

Since the prototypes were designed to periodically read energy data from household smart meters, a Python script was created that replicated this process. A series of data points, consisting of a pair of values for resp. energy consumption and energy production was modelled with this script serving as the main input for the prototype's feedback algorithm, see Figure 4.

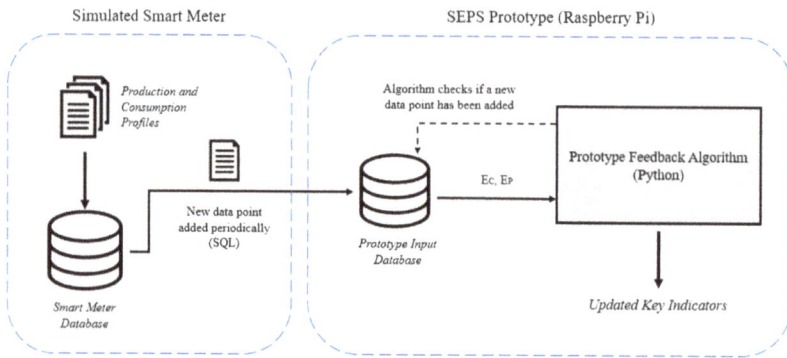

Figure 4. Test set-up for the scenario simulation testing.

2.3. End User Testing

In order to compare the results from scenario simulations to a real-life situation, two of the prototypes were briefly tested with end users in several households in the Netherlands. The tests were conducted in two phases:

● **Phase 1—Reference Measurements**

This phase was used to create a benchmark for evaluating the effectiveness of each prototype during the second phase. Household energy production and consumption were measured on 15-min intervals by connecting a Raspberry Pi unit directly to the household's smart meter.

● **Phase 2—HEMP Prototype Testing**

In this phase, users were presented with a brief description of the prototypes as well as a short demonstration of their operation, after which one of the prototypes was installed in their home. Users were then left to freely interact with the prototype for several days; during this phase there was constant monitoring of energy consumption and generation with the prototypes capturing data from smart meters at 15-min intervals.

3. Results

3.1. Simulation Environment Test Results

3.1.1. Bodhi

Figure 5 shows how the prototype's lighting reacted to a gradual increase in cumulative energy consumption relative to an arbitrary energy budget, going from the under budget state (left) to the on budget (centre) and over budget (right) feedback states.

Figure 5. Time-lapse showing Bodhi's lighting transitions through all three feedback states.

3.1.2. CrystalLight

Figure 6 shows different stages of the modelled charge-discharge cycle, where the prototype's LED brightness gradually increased before reaching its maximum intensity level, then becoming dimmer until the full discharge state was attained.

Figure 6. Time-lapse showing CrystalLight at different stages of a charge-discharge cycle.

3.1.3. LightInsight

The four system states for this HEMP were tested by first increasing the value of R_E from 0.9 to 1.1 (Figure 7, pictures 1–3) and later decreasing it (Figure 7, pictures 3–5) back to its initial value.

Figure 7. Time-lapse showing each of LightInsight's feedback states.

3.2. Scenario Simulation Results

3.2.1. Bodhi

This prototype was tested in two scenarios, corresponding to summer and winter loads since its operation is not dependent on PV production. In the summer scenario, a smooth transition through all three feedback states was observed, with energy consumption starting significantly under budget ($R_B < 0.95$) and staying on budget for a short interval before remaining consistently over budget ($R_B > 1.05$) for the rest of the day. This is partly due to the shape of the budget ratio curve itself, which shows an upward trend with short intervals where R_B sharply increases. These intervals match peaks in household load and are followed by gradual decreases as energy use reverts back to the baseline load. The selected energy budget (B) also had a significant impact on when these transitions took place since it determines the balance point between actual and planned consumption ($R_B = 1$).

The winter scenario showed a similar trend for the budget ratio throughout the day while showing more pronounced increases in R_B than during the summer scenario, as seen in Figure 8 below. Once again, the energy budget was exceeded by the end of the day, although this occurred much later; as was the case before, this greatly depended on the selected energy budget.

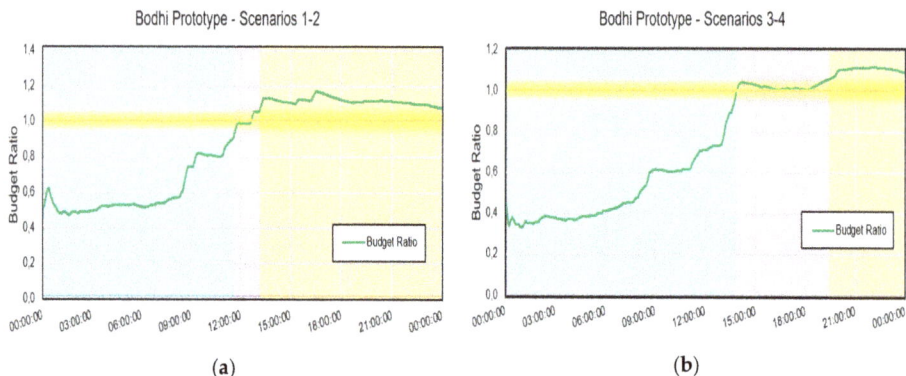

Figure 8. Bodhi prototype performance with: (**a**) a summer load profile (scenarios 1 and 2) and (**b**) a winter load profile (scenarios 3 and 4). Background colour in the figures corresponds to the light colour shown by the prototype LEDs; the yellow line indicates the balance point between actual and planned consumption.

3.2.2. CrystalLight

During Scenario 1, this prototype spent the vast majority of the day at full discharge, only charging during a few short intervals between 7:30 and 11:00 where the maximum charge, set at 15 Wh, was quickly reached and then consumed. This should not be surprising considering that energy consumption consistently outperforms production in this scenario.

In Scenario 2, fast charging took place from 7:00 to 10:00, with the prototype fully charged for around five hours before gradually discharging for the rest of the day as seen in Figure 9a). As expected, performance was significantly better than in the previous scenario; the only times in which a full discharge occurred were the early morning hours where PV production had not yet started.

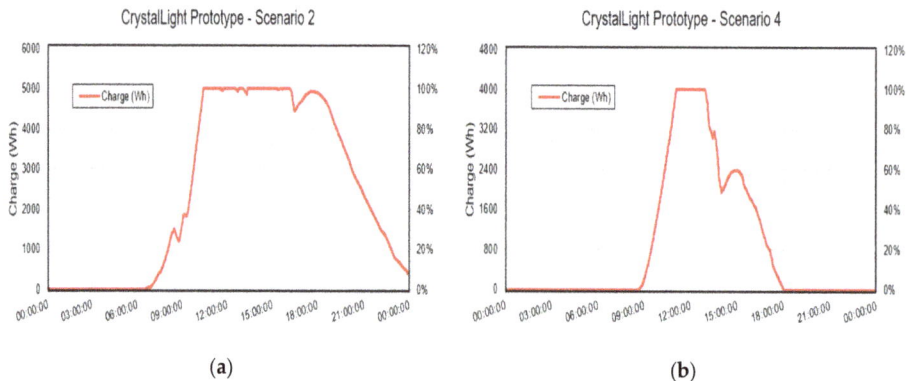

Figure 9. CrystalLight prototype performance during: (**a**) Scenario 2, and (**b**) Scenario 4. LED intensity, corresponding to the prototype's state of charge, is shown on the right.

The combination of poor PV production and high energy demand in Scenario 3 resulted in the prototype being fully discharged for the entire day; this means that from the user's perspective the product lights would be constantly off.

Finally, in a similar way to Scenario 2, in Scenario 4 the prototype went through a charge-discharge cycle during the daytime, with a second, shorter charging phase in the early afternoon (see Figure 9b below). The discharge phases were faster in this case, with the battery emptying completely by 18:00.

Maximum charge was set at 4000 Wh, which explains why the charging phase abruptly stopped at around 11:00.

3.2.3. LightInsight

The prototype's "net consumption" state took place around 93% of the time in Scenario 1, the only exception being several short periods of "net production" between 6:00 and 13:00 as seen in Figure 10a below. The two proposed transition states (corresponding to "rainbow" and "yellow" LED lighting) were extremely rare, each occurring less than 1% of the time. This is due to the abrupt changes observed for R_E, which hardly fell within the transition range ($0.95 < R_E < 1.05$).

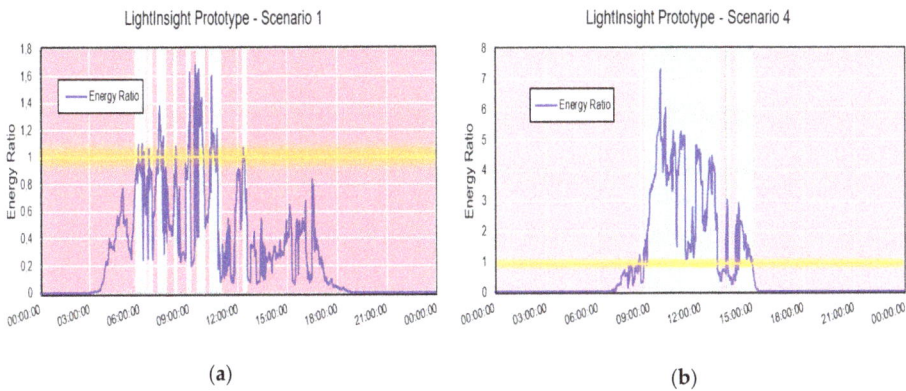

(a) (b)

Figure 10. LightInsight prototype performance during: (**a**) Scenario 1 and (**b**) Scenario 4. Background colour corresponds to the light colour shown by the prototype LEDs; the yellow line indicates the balance point between energy consumption and production.

As expected from the increased PV production in Scenario 2, net production periods were much more frequent, amounting to around 40% of the total intervals and lasting longer on average. The energy ratio was also significantly higher both on average ($R_E = 1.7$ compared to 0.3 from Scenario 1) and on its maximum value, exceeding $R_E = 10$ on several occasions. Transition states occurred even less frequently than in Scenario 1, both accounting for only 0.9% of the total intervals.

The performance of this prototype in Scenario 3 matched the observations made for CrystalLight since the low values of R_E failed to approach the transition range and red lights showed for the entire day. In Scenario 4, however, there were a few hours around noon where the net production state occurred with little to no interruption (see Figure 10b below). Transition states were less frequent than in any other scenario, with only two yellow intervals (0.14%) and one rainbow interval (0.07%) during the entire day. The prototype performed better than in Scenario 3 as expected but a better performance than in Scenario 1 was also achieved, showing that good PV production seems to have a more significant impact in this prototype's feedback than changes in household consumption.

3.3. End User Testing Results

3.3.1. Bodhi

The performance of this HEMP revealed that the selected energy budget, which was based on the household's average consumption during the previous week, greatly overestimated the actual energy use during testing. Consumption during the morning of the second day was much higher than expected but then sharply decreased, transitioning through all three feedback states and remaining on the under budget state for the rest of the day and the next two full days as well. Due to the short length of the testing period, it is hard to determine whether this decrease in consumption can be attributed to the users' reaction to the prototype or if there was influence from other factors.

Despite this overestimation, it is still possible to see that most daily budget ratio profiles follow a similar pattern consisting of an overall increasing trend with short intervals where R_B increases sharply. As was the case with the profiles observed in the scenario simulations, these intervals match peaks in household load and are followed by smaller gradual decreases as energy use reverts back to the baseline load. A noticeable exception to this pattern was observed in the first half of the second testing day where a decreasing trend took place, as shown in Figure 11.

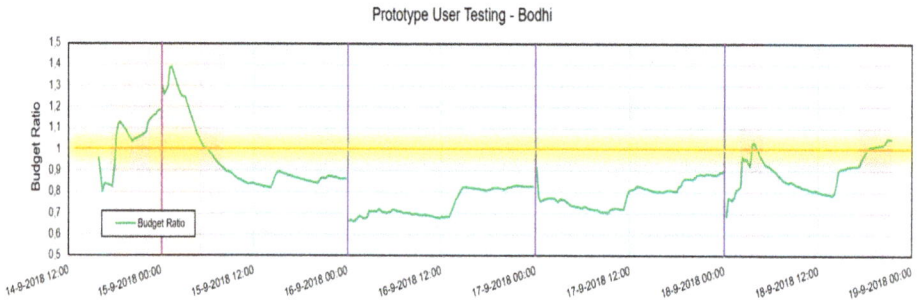

Figure 11. Bodhi prototype performance during testing phase. Background colour corresponds to the light colour shown by the prototype LEDs; the yellow line indicates the balance point between actual and planned energy consumption.

3.3.2. LightInsight

The performance of the LightInsight prototype during user testing showed strong similarities to the patterns observed in some of the simulated scenarios, with drastic changes to R_E taking place in brief periods of time (see Figure 12 below). Peaks are significantly more pronounced in some days than in others, possibly indicating sunny or overcast weather.

The prototype showed red LED lighting for most of the time, with scattered periods of net production appearing mostly during midday and the early afternoon (11:00–17:00). The net consumption state constituted around 86% of the total intervals; the yellow and rainbow transition states, at 1 (0.2%) and 2 (0.5%) intervals, almost never occurred during testing.

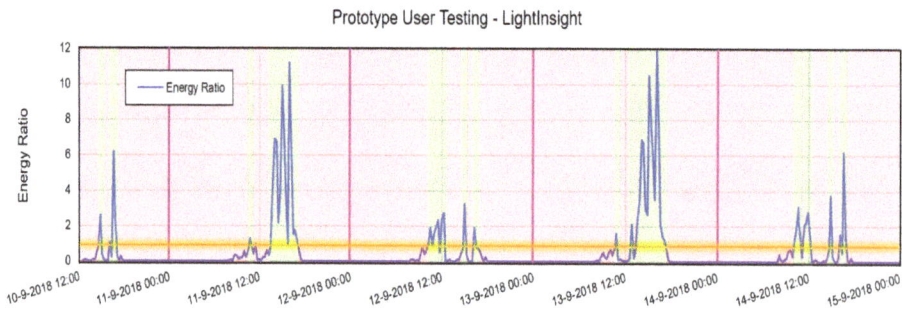

Figure 12. LightInsight performance during testing phase. Background colour corresponds to the light colour shown by the prototype LEDs; the yellow line indicates the balance point between energy consumption and production.

4. Discussion and Conclusions

The operation of the three HEMP prototypes was successfully tested using a simulation environment, proving the usefulness of this tool for quickly and accurately validating the operation of SEPS using simulated PV production and load profiles. The tests presented a simple, quick visualisation of how these prototypes would operate in households without the need to involve the end users themselves. This approach can be useful during the early product development phase for rapidly testing several modes of operation and determining which one is best suited for achieving the intended purpose of a given SEPS.

Although the proposed testing sequences were relatively simple, they provided a clear demonstration of how the HEMPs would operate in practice, and there is potential for improving the accuracy of these tests by designing more complex testing sequences. Future simulation testing of the presented HEMP concepts should also explore the potential for incorporating other methods such as co-simulation in order to obtain more accurate results. These tests would require the development of an agent-based model of product influence on user behaviour and the resulting impact on the residual load profile in order to upscale the physical device to a large number of simulated devices.

The scenario simulations and end user testing served as a more extensive test on HEMP performance, which helped identify some of the advantages and limitations of the current designs. For instance, tests on the Bodhi concept helped identify a recurring daily R_B profile but were limited by inaccurate predictions for the household's energy budget, highlighting the importance of correctly estimating this type of parameter during its operation. LightInsight, on the other hand, was able to present simple, intuitive information about how energy flows in a household, but was found to be useful for only a small fraction of the day since by definition net consumption takes place whenever the sun is down. The rapid fluctuations observed in the energy profiles also meant that the proposed transition states were extremely rare, meaning they should be restructured to better respond to the observed user behaviour. Overall, the proposed feedback algorithms required only basic energy data in order to determine feedback to users. However, developing more complex algorithms that use other variables as inputs and respond to changes in use patterns could help improve the effectiveness of these concepts in the future or even add new functions, such as the scheduling algorithm for smart appliances presented in [27].

The scenario-based simulations were also intended to replicate the conditions observed during the end user testing. User test performance for the LightInsight concept was more closely approximated by Scenario 1 with a root mean square error (RMSE) between both datasets of 1.58 followed by Scenarios 3, 4 and 2 (RMSE = 1.61, 1.96 and 2.43, respectively). For Bodhi, the summer scenario (RMSE = 0.31) matched the user tests more closely than the winter scenario (RMSE = 0.35). Overall, while the simulations approximated some of the general trends observed in the user tests, they are still far from being a significant predictor of SEPS operation. Comparing HEMP performance during user tests to their reference measurements, on the other hand, revealed that the concepts did not always seem to achieve their intended purpose. Testing on LightInsight resulted in an increase in both the average load (25%) and the peak load (3%) compared to the reference phase, and a more deficient match between energy supply and demand. User tests with Bodhi, on the other hand, had a positive impact since the overall energy consumption showed a significant decrease averaging 27% less than in the reference phase. Thus, these preliminary results can show at a glance whether a given SEPS concept is working adequately, and this information can be used to modify its design during the next development phase.

It is important to consider that the simulated scenarios covered a short period of time due to limited data availability; the use of larger datasets for a longer period of time could provide a more accurate representation of the modelled operating conditions. A study by van Dam et al. [11], which sought to evaluate the effectiveness of HEMS in saving energy in households, reinforces this notion since it had a much larger sample size and a longer duration than the present work. However, the study still indicated the need for conducting more extensive research into long-term effects, and also pointed out that the initial effectiveness of HEMS feedback tends to wear off with time.

Overall, the presented testing methods were successful in evaluating the potential of HEMP concepts and identifying possible challenges or bottlenecks in their design, offering valuable insights that can result in significant early improvements and make the product development process more efficient. The presented simulation environment can be a good first approach for testing and showcasing the operation of SEPS designs in a shorter period of time than with real user tests, with the possibility of obtaining more accurate results in the future through a more extensive co-simulation approach. In addition to this, scenario-based simulations using existing energy profiles can provide an accurate approximation of real-life conditions, revealing flaws or limitations which would otherwise come to light later on and would become much more difficult to overcome. Since only a limited number and a specific type of SEPS were tested in the present work, it is recommended that these tests are repeated with a wider range of SEPS concepts as well as with different energy system configurations to enhance the robustness of these methods.

Author Contributions: Conceptualisation, A.S. and A.R.; methodology and software, A.S. and S.Ü.; investigation and resources, C.G. and S.Ü.; data curation, A.S. and S.Ü.; formal analysis, visualisation and writing—original draft preparation, A.S.; writing—review and editing, C.G., S.U. and A.R.; supervision, project administration and funding acquisition, A.R.

Funding: This research has received funding from the European Union's Horizon 2020 research and innovation programme under the ERA-Net Smart Grids plus, grant number 646039, from the Netherlands Organisation for Scientific Research (NWO). This work was also partially funded by the European Commission within the H2020 framework ERIGrid project under Grant Agreement No. 654113.

Acknowledgments: The content and views expressed in this material are those of the authors and do not necessarily reflect the views or opinion of the ERA-Net SG+ initiative. Any reference given does not necessarily imply the endorsement by ERA-Net SG+. We would also like to thank David Reihs for his technical support in conducting the simulation tests at AIT.

Conflicts of Interest: The authors declare no conflict of interest.

References

1. Reinders, A.; De Respinis, M.; Van Loon, J.; Stekelenburg, A.; Bliek, F.; Schram, W.; van Sark, W.; Esteri, T.; Übermasser, S.; Lehfuss, F.; et al. Co-evolution of smart energy products and services: A novel approach towards smart grids. In Proceedings of the Asian Conference on Energy, Power and Transportation Electrification, ACEPT 2016, Singapore, 25–27 October.

2. Lannoye, E.; Flynn, D.; O'Malley, M. Evaluation of Power System Flexibility. *IEEE Trans. Power Syst.* **2012**, *27*, 922–931. [CrossRef]

3. Smale, R.; van Vliet, B.; Spaargaren, G. When social practices meet smart grids: Flexibility, grid management, and domestic consumption in The Netherlands. *Energy Res. Soc. Sci.* **2017**, *34*, 132–140. [CrossRef]

4. Gercek, C.; Reinders, A. Smart Appliances for Efficient Integration of Solar Energy: A Dutch Case Study of a Residential Smart Grid Pilot. *Appl. Sci.* **2019**, *9*, 581. [CrossRef]

5. Zhang, Z.; Gercek, C.; Renner, H.; Reinders, A.; Fickert, L. Resonance Instability of Photovoltaic E-Bike Charging Stations: Control Parameters Analysis, Modeling and Experiment. *Appl. Sci.* **2019**, *9*, 252. [CrossRef]

6. Reinders, A.; Übermasser, S.; Van Sark, W.; Gercek, C.; Schram, W.; Obinna, U.; Lehfuss, F.; Van Mierlo, B.; Robledo, C.; Van Wijk, A. An Exploration of the Three-Layer Model Including Stakeholders, Markets and Technologies for Assessments of Residential Smart Grids. *Appl. Sci.* **2018**, *8*, 2363. [CrossRef]

7. Robledo, C.B.; Oldenbroek, V.; Abbruzzese, F.; van Wijk, A.J.M. Integrating a hydrogen fuel cell electric vehicle with vehicle-to-grid technology, photovoltaic power and a residential building. *Appl. Energy* **2018**, *215*, 615–629. [CrossRef]

8. Mwasilu, F.; Justo, J.J.; Kim, E.; Do, T.D.; Jung, J. Electric vehicles and smart grid interaction: A review on vehicle to grid and renewable energy sources integration. *Renew. Sustain. Energy Rev.* **2014**, *34*, 501–516. [CrossRef]

9. Schram, W.L.; Lampropoulos, I.; van Sark, W.G.J.H.M. Photovoltaic systems coupled with batteries that are optimally sized for household self-consumption: Assessment of peak shaving potential. *Appl. Energy* **2018**, *223*, 69–81. [CrossRef]

10. Posma, J.; Lampropoulos, I.; Schram, W.; van Sark, W. Provision of Ancillary Services from an Aggregated Portfolio of Residential Heat Pumps on the Dutch Frequency Containment Reserve Market. *Appl. Sci.* **2019**, *9*, 590. [CrossRef]

11. Van Dam, S.; Bakker, C.; van Hal, J. Home energy monitors: Impact over the medium-term. *Build. Res. Inf.* **2010**, *38*, 458–469. [CrossRef]

12. Geelen, D.; Reinders, A.; Keyson, D. Empowering the end-user in smart grids: Recommendations for the design of products and services. *Energy Policy* **2013**, *61*, 151–161. [CrossRef]

13. Hargreaves, T.; Nye, M.; Burgess, J. Making energy visible: A qualitative field study of how householders interact with feedback from smart energy monitors. *Energy Policy* **2010**, *38*, 6111–6119. [CrossRef]

14. Van Mierlo, B. Users Empowered in Smart Grid Development? Assumptions and Up-To-Date Knowledge. *Appl. Sci.* **2019**, *9*, 815. [CrossRef]

15. Obinna, U.; Joore, P.; Wauben, L.; Reinders, A. Insights from Stakeholders of Five Residential Smart Grid Pilot Projects in the Netherlands. *Smart Grid Renew. Energy* **2016**, *7*, 1–15. [CrossRef]

16. Wolsink, M. The research agenda on social acceptance of distributed generation in smart grids: Renewable as common pool resources. *Renew. Sustain. Energy Rev.* **2012**, *16*, 822–835. [CrossRef]

17. Palensky, P.; van der Meer, A.A.; Lopez, C.D.; Joseph, A.; Pan, K. Cosimulation of intelligent power systems. *IEEE Ind. Electron. Mag.* **2017**, *11*, 34–50. [CrossRef]

18. Godfrey, T.; Mullen, S.; Dugan, R.C.; Rodine, C.; Griffith, D.W.; Golmie, N. Modeling Smart Grid Applications with Co-Simulation. In Proceedings of the 2010 First IEEE International Conference on Smart Grid Communications, Gaithersburg, MD, USA, 4–6 October 2010.

19. Faruque, M.O.; Sloderbeck, M.; Steurer, M.; Dinavahi, V. Thermoelectric co-simulation on geographically distributed real-time simulators. In Proceedings of the IEEE PES General Meeting, Calgary, AB, Canada, 26–30 July 2009; pp. 1–7.

20. Georg, H.; Muller, S.; Rehtanz, C.; Wietfeld, C. Analyzing cyber-physical energy systems: The INSPIRE cosimulation of power and ICT systems using HLA. *IEEE Trans. Ind. Inf.* **2013**, *10*, 2364–2373. [CrossRef]

21. Khan, M.; Silva, B.N.; Han, K. Internet of Things Based Energy Aware Smart Home Control System. *IEEE Access.* **2016**, *4*, 7556–7566. [CrossRef]

22. Guenther, C.; Schott, B.; Hennings, W.; Waldowski, P.; Danzer, M.A. Model-based investigation of electric vehicle battery aging by means of vehicle-to-grid scenario simulations. *J. Power Sour.* **2013**, *239*, 604–610. [CrossRef]

23. Vardakas, J.S.; Zorba, N.; Verikoukis, C.V. Performance evaluation of power demand scheduling scenarios in a smart grid environment. *Appl. Energy* **2015**, *142*, 164–178. [CrossRef]

24. Liedtke, C.; Baedeker, C.; Hasselkuss, M.; Rohn, H.; Grinewitschus, V. User-integrated innovation in Sustainable LivingLabs: An experimental infrastructure for researching and developing sustainable product service systems. *J. Clean. Prod.* **2015**, *97*, 106–116. [CrossRef]

25. Ceschin, F. Critical factors for implementing and diffusing sustainable product-Service systems: Insights from innovation studies and companies' experiences. *J. Clean. Prod.* **2013**, *45*, 74–88. [CrossRef]

26. AIT SmartEST Laboratory for Smart Grids (Fact Sheet). Available online: https://www.ait.ac.at/fileadmin/mc/energy/downloads/Smart_Grids/Produktblatt_CI_SmartEST_lowRes.pdf (accessed on 2 April 2019).

27. Chavali, P.; Yang, P.; Nehorai, A. A Distributed Algorithm of Appliance Scheduling for Home Energy Management System. *IEEE Trans. Smart Grid* **2014**, *5*, 282–290. [CrossRef]

Review

Users Empowered in Smart Grid Development? Assumptions and Up-To-Date Knowledge

Barbara van Mierlo

Wageningen University & Research, Knowledge, Technology and Innovation, P.O. Box 8130, 6700 EW Wageningen, The Netherlands; barbara.vanmierlo@wur.nl; Tel.: +31-317483258

Received: 19 November 2018; Accepted: 19 February 2019; Published: 26 February 2019

Abstract: Active involvement of users in smart grids is often seen as key to beneficial development of smart grids. In this paper, we investigate the diverse assumptions about how and why users should be active and to what extent these assumptions are supported by experiences in practice. We present the findings of a systematic literature review on four distinctive forms of user involvement in actual smart grid projects: demand shifting, energy saving, co-design, and co-provision. The state-of-the-art knowledge reflects the preoccupation with demand shifting in the actual smart grid development. Little is known about the other user roles. More diversity in types of projects regarding user roles would improve the knowledge base for important decisions defining the future of smart grids.

Keywords: smart grids; users; demand management; renewable energy transition

1. Introduction

Active involvement of users in smart grids is often seen as key to beneficial development of smart grids. This is signaled, among other things, by the abundant use of the term "prosumer" and similar terms, such as "co-provider", in the context of smart grids. It can also be seen in some of the definitions of smart grids that stress the role of users, such as the following: "A Smart Grid is an electricity network that can cost efficiently integrate the behaviour and actions of all users connected to it—generators, consumers and those that do both—in order to ensure economically efficient, sustainable power system with low losses and high levels of quality and security of supply and safety" [1] (p. 2).

Energy companies and policy institutions are developing methods to involve the users. The European Commission, for instance, pursues a so-called "user-centric" approach, involving an increased interest in electricity market opportunities, value added services, flexible demand for energy, lower prices, and microgeneration opportunities [2]. Scholars have emphasized the importance of active engagement of users, among other things, by stating that they should be an empowered part of the system [3]. It is expected that if users are empowered, smart grids may become an important element in climate change policy and the renewable energy transition.

Active involvement of users promoting smart grids is, however, not self-evident. Demonstrations against smart meters in the UK and the USA have shown that users have specific concerns and may actively resist smart developments [4]. In early accounts, several social scientists have warned against the dominant technological focus on the development of smart grids [5,6]. Wolsink [6] stated that smart grids may not further the deployment of distributed renewables due to a lack of understanding of how and why users would accept smart grids. He sketched two scenarios. In the first, groups of end users deploy renewable energy sources in microgrids and increase their autonomy in relation to central power suppliers. In this scenario, the common resources, i.e., the locally produced renewable electricity, are deployed optimally. In the second, grid operators gain power by increasing surveillance of domestic consumers and their energy consumption. In this scenario, demand regulation is the main driving force rather than the optimal use of renewable energy. This one may gain dominance as it fits better with dominant relationships in the energy sector and the common, centralized patterns of operating. An inventory of the first developments in practice, i.e., the application of smart grids in the first pilot projects, showed that social aspects have received little attention [7]. There is a strong engineering bias with a focus on new information and communication technologies (ICT), disregarding their interaction with other technological and social aspects of local energy systems. Moreover, economic rationales and automation are central elements of smart grid project designs, even though the projects vary considerably [8]. The roles and interactions of users with other stakeholders, market, and technology development should be evaluated for a solid understanding of smart grid development, especially in the residential sector [9].

There seems to be a contrast between the terminology implying active roles of users and the actual developments. As Schick [10] (p. 82) says, "Even though users play an important role in the imagination of experts, a gap remains between the experts and those who smart electricity infrastructures will come to affect". Apparently, different assumptions about how and why users should be involved in smart grid developments co-exist. This raises the questions of what these assumptions are and to what extent these are supported by the current experiences with smart grids in practice. As a first, crucial step to answering these questions, we define the state-of-the-art knowledge on these issues on the basis of a literature review.

In the following sections, we first present the research methodology. Subsequently, the results of the literature review are discussed with regard to four types of active user roles: demand shifting, energy saving, co-design, and co-provision. The paper ends with a discussion on important topics for further research and the implications of up-to-date knowledge for smart energy products and services (SEPS) as well as smart grid projects.

2. Methodology

In the past years, the early employment of smart grid in countries such as Denmark, the UK, and the Netherlands has been investigated in social scientific studies. These provide the foundation of the current knowledge about the roles of users and other stakeholders. The literature is scattered because most studies are (comparative) case studies, investigating one or two specific smart grid projects. As a consequence, an overview of the state-of-the-art knowledge is missing. In this paper, we present the findings of a systematic literature review, analyzing (1) the main assumptions regarding user involvement and (2) the lessons that can be derived about these assumptions from practice.

First, the scientific literature on user involvement in real-life smart grid developments was collected using the search term "smart grid" and synonyms thereof, such as "smart home", on the one hand and terms related to users on the other hand ("user", "prosumer", "customer", "consumer", "co-provider"). In this way, articles to be included in the analysis were selected from the web-based Google Scholar database. To find missing literature on community-led initiatives, we had a second round of search for "smart grids" in combination with "local energy initiatives" or related terms, such as "energy cooperative", "autonomy", "independence", and "micro-grid".

Secondly, to select the most relevant articles for the analysis, we screened the abstracts and methodology sections on the following criteria: (1) social scientific research, (2) peer-reviewed article in international journal, (3) relevant empirical evidence of actual experiences with smart grids, and (4) sound research methodology according to the standards in the diverse research domains. The resulting selection of articles was mostly about studies of smart grid pilot projects. Surveys among prospective users were included only if they provided additional insights. Ultimately, 43 studies were included in the analysis; some of these were studies were by the same authors.

Thirdly, the selected articles were roughly screened to see what the perspective on user involvement was. We classified types of user involvement in four distinct user roles.

Finally, the selected articles were organized by the four defined user roles, after which the findings and conclusions of the selected articles were qualitatively analyzed as secondary data. For each user role, the knowledge about the involvement of users in smart grids was synthesized as well as categorized on key topics. As a consequence, not all selected literature is mentioned in this paper.

The smart grid definition mentioned above stresses its technological characteristics and prioritizes cost-efficiency over other goals; it is even mentioned twice in the definition. Many other definitions emphasize digital technology and the communication through ICT-based devices. In our study, in order to develop a wide perspective on user roles, we followed a smart grid definition that includes the social features: "A smart grid is a socio-technical network characterized by the active management of both information and energy flows, in order to control practices of distributed generation, storage, consumption and flexible demand" [6] (p. 824).

The goals of the research were as follows:

1. To uncover the assumptions about user roles, motivations, responses, and energy behavior in scientific literature about smart grids and user involvement. What kind of involvement is considered key for which goals related to smart grids and with what argument?
2. To evaluate these assumptions in the light of the practice of smart grid development. To what extent are these assumptions confirmed in smart grid projects?
3. To evaluate the current knowledge. What is the state-of-the-art knowledge regarding the assumptions about user involvement in smart grids and what are the gaps therein?
4. To explore the implications of the findings for smart grid development. What conclusions can be derived from the findings for the need for further research and development of smart grids products, services, and projects?

3. Results

The assumptions about the roles of users differed throughout the literature depending on the key topic addressed. The topics were in turn often related to the specificities of the pilot project. The most important knowledge about each user role is presented herein by discussing the dominant assumptions and relating them to the key findings in recent empirical research.

3.1. Demand Shifting

The literature on the topic of flexibility in relation to smart grids ties in with the older debates on demand-side management. The key challenge is to match supply and demand better in order to accommodate the distributed energy sources and reduce CO_2 emissions. Users are expected to respond to information about the availability of electricity or financial incentives, such as dynamic pricing, by shifting the energy load to preferred moments or a combination thereof. They tend to be addressed as isolated individuals making their own, autonomous decisions. The implicit assumption underlying information services is that new knowledge and awareness might motivate users to shift their energy use for environmental or financial reasons.

However, there were several examples of "disappointing" responses in the investigated projects. In the Jouw Energie Moment (Your Energy Moment) project, for instance, householders had requested for "smarting" their electrical heat pumps, but only four out of 38 actually switched it on [11].

Since 2014, several good social scientific studies have been conducted on demand shifting. Most of them related to what is called the "practice turn" in social sciences. The idea is that genuine interest in what moves people in their homes to consume energy helps to understand to what extent and under what conditions they may shift their energy consumption to other moments of the day.

A first insight is that domestic practices are pinned in time and place, related to relationships within the households, social conventions, and time structures of the activities of members of the households [12,13]. As a consequence, some practices are more prone to active time shifting than others. In general, cleaning practices (dishwashing, washing, and tumble drying) were found to be most suitable for demand-side response [14,15]. Practices implied in ambiance regulation, leisure, cooking, and eating, are less easy to change. Solitary tasks are easier to change than collective ones, such as family dining as well as practices that depend on the structure of activities outside the home.

Secondly, several authors confirmed that households have to "learn" to adapt their demand, i.e., change their practices and start seeing options for doing so [8,12,16]. In two projects investigated by Hansen [8], the knowledge about household consumption, electricity markets, prices, and electricity system loads had increased. There were, however, also accounts of what could be called "unlearning" the practices geared toward demand shifting. Kessels [16], for instance, showed that there was a "response fatigue" in one of the investigated projects in case of manual feedback and control.

Thirdly, flexibility interventions were seen to influence relationships among household members and to even have undesirable effects. Skjølsvold et al. [17] emphasized internal household dynamics between men and women. Drawing on two Norwegian smart grid demonstration projects, the provided feedback was found to "trigger" learning amongst eager men while alienating or excluding women.

Three key issues in relation to demand shifting were addressed in the literature: (1) automation and control, (2) financial incentives, and (3) feedback and communication.

(1) A key issue in demand shifting is the level of automation and control. Several authors emphasized that automation defines demand shifting more than economic incentives [8]. While earlier studies raised concerns about a loss of control, more recent studies provided another image [18,19]. In general, remote or automatic control seems to be acceptable to residents, that is, under conditions. It depends on what equipment is controlled, the information and security of smart techniques provided, and whether they have the ability to override the external control [8,19]. Results of a representative survey indicated that a direct load control tariff was more acceptable than the time-of-use tariffs presented [18]. The load control tariff gave residents a better sense of control over comfort, timing of activities, and spending as well as ease of use. The majority of respondents were inclined to accept direct load control if they would have the option to override it. Another study on actual projects showed similar results. The households involved did not mind having their heating devices (heat pumps and electric heating) remotely controlled, but they did not appreciate automatic charging of electric vehicles and automated control of other devices, such as freezers, fridges, and pumps [8].

Interestingly, even in the case of full automation, residents may show active engagement, for instance, by connecting extra devices to the system [8]. In general, however, active engagement by members of households is expected to involve only the anticipated responses to informational and financial incentives.

(2) Flexible network tariffs are based on the assumption that energy consumption is financially motivated. In an important study based on a meta-review of literature and an empirical validation in 32 European projects with user engagement, the following was concluded [16]:

- Time of use tariffs have more potential than dynamic tariffs (real-time [ricing), while the latter is more relevant in the case of local energy production.
- In order to work effectively, the dynamic tariff

    - should be simple to understand for the end users,
    - should have timely notifications of price changes, and
    - should have a considerable effect on their energy bill.

If the tariff is more complex, the burden for the consumer could be eased by introducing automated control.

Whether dynamic tariffs will be accepted is most likely dependent on whether users consider them to be fair. A recent study [19] found that transport or capacity charges are considered fairer than peak pricing and much more than a flat rate. What is considered fair depends on noneconomic justification, guarantee that basic needs will be fulfilled, predictability, and being sure that peak use is not only affordable for rich people.

Several studies indicated that there is seldom a linear relationship between financial incentives and energy consumption. This is explained by the intermediating role of domestic practices, as mentioned above. It also relates to the fact that considerations other than financial ones influence energy consumption. Saving energy, an interest in new techniques, and a green image are among the other reasons for becoming flexible [5,19]. If households are primarily motivated by a desire to become self-sufficient and autonomous, this could even create a tension with demand-side flexibility schemes, which tend to create new types of grid dependency (whether it be flexibility contracts, automated demand response, or market-based dynamic energy pricing incentives). In such cases, users may prefer to stick to established forms of "green engagement" with energy, such as green energy contracts and applying energy-efficient light bulbs [14].

(3) A recent study confirmed findings from earlier studies regarding the effectiveness of feedback provided [8]. Simple information and visualization methods, such as a light signal (green/red) and emails or text messages that integrate information about demand and its relationship with supply, tend to have a positive influence (see also [3]). Users evaluate them positively, understand the information, and tend to change their consumption patterns based on the input.

More complex interfaces provided ambiguous results. Web-based tools were used by less people: approximately 25% in 11 Danish experiments [8]. Moreover, the satisfaction with the applicability of websites and advanced in-home boxes varied. Kendel [20] showed that if people are willing to visit a portal, more advanced information per appliance could slightly increase the effectiveness in flexibility.

It tends to be forgotten, however, that people's awareness of patterns of energy consumption is not only provided by direct, individual feedback. Naus [12] nicely showed how, in addition to individual learning, community interaction in the form of workshops and informal encounters with neighbors contributes to effective demand shifting.

In conclusion, it can be stated that the insights about demand shifting have considerably increased in the past years due to a large number of studies. It has been shown that there is certainly some room for triggering active responses from users, even though the relationship between incentive and energy consumption is mediated by practices and motives other than that intended with the measure. The findings of the studies provide some clear rules of thumb. However, whether and how measures for automation, financial incentives, and feedback mechanisms are effective depends ultimately on the specific features of the local energy system and the local setting, such as the relationships between residents and the project managers, conventions embedded in culture, weather conditions, etc. This was well illustrated in the comparative study by Bulkeley [21], which showed the outcomes of different settings and interventions. The first group of users had photovoltaic (PV) panels and a display showing their domestic consumption. This group was stimulated to do financial calculations of the revenues of possessing PV. The second group had PV and a device showing the moments of available electricity. The users were stimulated to change their routines of washing and dishwashing (modestly), among other things, by a timer on the washing machine. They were also engaged in energy management by checking weather forecasts. The third group was provided with PV and a water tank that automatically absorbs excess generation. This group started shifting their showering practices from the morning to the daytime or evening—a response that is rather surprising in the light of the social practices studies mentioned above.

3.2. Energy Saving

Smart grid systems are expected to stimulate residential users to save electricity. This expectation stems from two integrated smart grid developments. First, the diffusion of distributed renewable energy may trigger a motivation to save energy. Here, the mere possession of renewable energy technologies is assumed to trigger further behavioral change. A literature review reported that little is known about this effect [22]. While the installation of a PV system stimulates many households to reduce their overall electricity consumption according to themselves, the rare reports on actual consumption show otherwise, partly due to rebound effects.

Secondly, smart meters are applied to provide residential users with regular information about their electricity consumption who hitherto lacked such insight. The related assumption is that energy consumption data, supplemented with, for instance, the related financial incentives, or a comparison with other households will stimulate a reduction in electricity consumption. The data provided with smart meters, however, are seldom the preferred and recommended real-time data.

For groups of users, energy saving seem to be quite important. Smale et al. [14], for instance, reported that not all discussants in the focus groups were convinced of the primacy of the time-shifting problem over other sustainability issues. Hansen and Borup [8] concluded on the basis of a comparison of 11 Danish experiments that other motivations are an interest in new energy technologies, being more environmentally friendly, and reducing the overall energy consumption.

Few studies have investigated energy saving in smart grid pilot projects. One of the exceptions is the Danish eFlex project. The study showed an overall reduction in electricity consumptions in response to the feedback from home energy management systems: "the 'control' household group, [with only the management system] . . . had on average saved approximately 10% on their kilowatt hour consumption during the project period March 2011 to February 2012" [15].

While energy saving is of importance for specific user groups, it has received too little attention to arrive at conclusions about the relationship between smart grids and energy saving. A complicating factor is that the term "energy saving" has become ambiguous with the introduction of smart grids. Some researchers use it for demand shifting rather than reducing the overall energy consumption (see, for instance, [20]). Moreover, due to the electrification trend accompanying the smart grid development, just assessing increases or decreases in kilowatt hour consumption has little meaning. The meaning of, and aspirations regarding energy saving hence needs reviewing in the context of smart grids.

3.3. Co-Design

Many social scientists and some engineers critique the technological and economic rationality in the design of smart grid technologies and projects. Hansen et al. [8] (p. 260), for instance, stated: "Our analysis shows that the projects employ a technology driven approach to household users with a focus on testing ready-made technologies rather than on improving technologies by including consumers". For this reason, several scholars presuppose that if the smart grid technologies are co-designed, they will better address the users' needs [3,7]. There were some examples where future users were addressed with images or prototypes of potential products. However, we could not find any study on future users purposively and structurally being involved in the preparation phase of pilot projects.

That does not mean that users play no role at all in design processes. In the literature, we found several examples of projects in which the applied techniques or their applications were changed because of users' wishes and responses [15,19]. Hansen and Hauge [19] showed how, in a project in which 20 households were to be equipped with the same air/water heat pump, it was installed in only seven households. After a negotiation process, the other households instead received a hybrid air/water heat pump with a gas boiler installed, sunwells, or a geothermal heat pump.

It can be concluded that co-design is rare and has so far yielded little insights in the preferences of users. The documentation on interactions between users and project managers in the course of actual projects provides evidence that, at times, feedback from users does influence choices made in these projects. Allowing room for such forms of co-design during the use phase of projects may instigate a countervailing power against the misconceptions of leading engineers.

In general, it seems useful to think of co-design not at the level of products and services only but at the level of projects as well. In this way, socio-technological aspects regarding the local energy system and stakeholder relationships will be part of the negotiations rather than merely the interaction between residents and technology.

3.4. Co-Provision

Where smart grids projects involve renewable energy, users are often regarded as co-providers. Instead of the passive consumer receiving energy, he/she is also producing and supplying energy. Such co-provision is regarded as an active role of users because their conduct "influences the grid and community, by reducing risks of load and voltage problems enabling more households to use PV" [21] (p. 17). Such perceptions of co-provision are merely technical; just the possession of renewable energy makes residents a provider. They are connected to the preoccupation with the flexibility dimension of smart grids. The related literature has been discussed above in the section "demand shifting".

A more radical perception is that of active, responsible citizens in the renewable energy transition process, where the relationships between the energy sector and users change drastically if users indeed take and get the responsibilities and power of being an energy provider. Transition advocates as well as several social scientists regard local energy initiatives (LEIs) to have the potential to transform the energy sector from fossil-fuel-based to renewable-energy-based by changing the relationships between citizens and institutional stakeholders. The related literature about LEIs forms a niche in the social scientific literature; smart grids are only touched upon as an issue. This may well mirror the priorities of the cooperatives.

A radical form of LEIs is the independent, local electricity systems. In microgrids or virtual power plants, prosumers may use, produce, and trade electricity without the interference of a central authority. The Brooklyn Microgrid is an example of an experiment in which energy is traded between prosumers and consumers. The grid operator has access to consumer data and manages energy use, loads, and demand response at negotiated rates. Quantitatively, the energy cooperatives do not yet contribute substantially. In the Netherlands, for instance, just 2% of the solar power installed was collectively owned in 2017 [23]. In this country, two cooperative energy suppliers organize 107 local energy initiatives, aiming to close the energy cycle. A study by Arentsen and Bellekom [24] illustrated that local energy associations often combine localized with centralized features, which is why the authors concluded that (p. 1) "local electricity initiatives can be considered a seedbed of innovation but with no potential to develop dominance in the electricity supply". Instead, the authors expected that the local initiatives will develop as niches inside the dominant electricity system, challenging its centrality and ever-increasing scale, and will add to the hybridization of its products and services. A dissimilar conclusion was drawn by Blanchet [25]. He studied the role of two LEIs in what he called the "remunicipalization" of Berlin's electricity grid and concluded that the (potential) impact of local initiatives on energy systems and their governance has been underestimated.

Unfortunately, there are few documented experiences with a truly active role of users as co-providers, probably due to demotivating policy. The findings are ambiguous regarding the potential of LEIs to change the energy system. Experiences in citizen-led initiatives may well provide very important insights on co-provision in addition to the industry- and government-led pilot projects that outnumber them.

4. Discussion

A need for user involvement in smart grids is often proclaimed. Users should be involved, among other things, for a good performance of smart grids, for the development of good SEPS, or a valuable contribution of this innovation to the renewable energy transition. The state-of-the-art knowledge on user involvement was identified here on the basis of a literature review of findings regarding four user roles.

Current knowledge predominantly concerns the user role of demand shifting. The results of the studies on this issue both confirm and contradict the assumptions about how users would respond to incentives. They provide some rules of thumb for triggering active responses. The results are somewhat scattered about subthemes due to the high diversity in energy systems in smart grid projects.

The studies steer away from the socio-psychological and economic theories that have dominated studies on energy users for decades. Sociological theories about social practices are shown to be fruitful to understand the complexity of changing practices involving energy use at the household or community level, especially regarding demand shifting.

On energy saving, the literature provides some limited dispersed knowledge. No valid conclusions regarding energy saving could be drawn. Nonetheless, it is a very important role in the eyes of users themselves and hence should be investigated further. Regarding co-design, it seems that users are hardly involved in the design of the technology and the projects, which means that knowledge on this type of involvement is lacking. Yet, there are indications of feedback loops from users to developers in running projects. Co-provision can be recognized in the emergence of LEIs. However, the findings on user involvement are ambiguous, and smart grids are not a priority of energy cooperatives.

An additional limitation of current research is that the relationships between users and stakeholders in the energy and other relevant sectors, such as ICT, are often disregarded. These relationships typically differ in the diverse pilot projects and influence user roles. Moreover, most studies that do focus on users do not investigate them in light of a wider renewable transition process and the role of smart grids therein. The question of whether and to what extent users are empowered remains unanswered. This may well be related to the optimism among smart grid advocates that smart grid transitioning itself guarantees a sustainable future.

Further empirical research evaluating the four user roles in current smart grid development is crucial. In such research, it is recommended to keep an open eye for unanticipated and unintended responses of users. This is very relevant in this early phase of smart grid introduction to society, which involves many uncertainties. Furthermore, it would be good to conduct further empirical research on standardization and smart grid policy and their influence on the decision-making process of the organizations initiating smart grid projects (particularly grid operators) and ultimately on user involvement.

A major implication of the findings for smart grid products and services is that they should take into account people's practices and relationships within the households and with their neighbors. Powells et al. [13] suggested to not only invent ways to let people be more flexible in the timing of their practices but to also reconfigure their practices in novel ways, especially those restricted by time constraints or social conventions. Slow cookers, for instance, would enable people to continue to eat together while shaving peak demand. Another proposed practice-aligned intervention is a launderette that uses stored, solar-heated hot water to clean clothes, which could provide services to both communities and those managing the distribution network.

The second implication is that the design and application of smart energy products and services should take place in an interactive process with (potential) users until there is sufficient evidence of what stimulates or hinders them to take on specific roles. Pilot projects as well as local energy initiatives provide the opportunity to regard SEPS not in an isolated way and users not as single, rational individuals but as part of a specific socio-technological system. Learning about smart grid development requires developing projects with a focus on local generation, local control, and cultural change in addition to the dominant projects aiming for enhanced efficiency and reliability. This would provide knowledge about the meaning of current socio-technological changes in the light of the contrasting future visions on smart grid development [6,25], in short, knowledge about the pathways of change regarding smart grids and distributed renewable energy.

5. Conclusions

While expectations of user involvement in smart grids are high, ideas about what it is and for what reasons it is important are ambiguous. The four user roles provide clarity regarding the assumptions about different types of user involvement. Evaluation of these assumptions on the basis of existing empirical studies shows that current knowledge about user involvement in smart grid development predominantly relates to the role of demand shifting. Knowledge on the roles of energy saving, co-design, and co-provision is limited.

Appl. Sci. **2019**, *9*, 815

The state-of-the-art knowledge exposes the preoccupation with demand shifting in the actual smart grid development. The expectations for user involvement reflected in terminology such as "empowered users" and "co-providers" is hardly mirrored in the practice of smart grid projects. More diversity in types of projects regarding user roles would improve the knowledge base for important decisions defining the future of smart grids.

Funding: This literature study is part of the interdisciplinary Co-Evolution of Smart Energy Product and Services (CESEPS) project. The project is funded in the framework of the joint programming initiative ERA-Net Smart Grids Plus, with support from the European Union's Horizon 2020 research and innovation programme.

Acknowledgments: I am grateful for the constructive comments of four anonymous reviewers. I also thank Esin Gültekin for the insightful discussions in the early preparation of this paper and Hilde Brouwers, Cihan Gercek, and Angèle Reinders for their feedback on the findings of the literature review.

Conflicts of Interest: The author declares no conflict of interest. The funders had no role in the design of the study; in the collection, analyses, or interpretation of data; in the writing of the manuscript; or in the decision to publish the results.

References

1. EC Directorate-General for Energy. *Standardization Mandate to European Standardisation Organisations (ESOs) to Support European Smart Grid Deployment*; M/490; European Commission: Brussels, Belgium, 2011.
2. European Technology Platform. *Vision and Strategy for Europe's Electricity Networks of the Future*; European Commission: Brussels, Belgium, 2006.
3. Obinna, U.P. *Assessing Residential Smart Grids Pilot Projects, Products and Services: Insights from Stakeholders, End-Users from a Design Perspective*; Delft University of Technology: Delft, The Netherlands, 2017.
4. Sovacool, B.K.; Kivimaa, P.; Hielscher, S.; Jenkins, K. Vulnerability and resistance in the United Kingdom's smart meter transition. *Energy Policy* **2017**, *109*, 767–781. [CrossRef]
5. Verbong, G.P.; Beemsterboer, S.; Sengers, F. Smart grids or smart users? Involving users in developing a low carbon electricity economy. *Energy Policy* **2013**, *52*, 117–125. [CrossRef]
6. Wolsink, M. The research agenda on social acceptance of distributed generation in smart grids: Renewable as common pool resources. *Renew. Sustain. Energy Rev.* **2012**, *16*, 822–835. [CrossRef]
7. Geelen, D.; Reinders, A.; Keyson, D. Empowering the end-user in smart grids: Recommendations for the design of products and services. *Energy Policy* **2013**, *61*, 151–161. [CrossRef]
8. Hansen, M.; Borup, M. Smart grids and households: How are household consumers represented in experimental projects? *Technol. Anal. Strateg. Manag.* **2018**, *30*, 255–267. [CrossRef]
9. Reinders, A.; Übermasser, S.; van Sark, W.; Gercek, C.; Schram, W.; Obinna, U.; Lehfuss, F.; van Mierlo, B.; Robledo, C.; van Wijk, A. An exploration of the three-layer model including stakeholders, markets and technologies for assessments of residential smart grids. *Appl. Sci.* **2018**, *8*, 2363. [CrossRef]
10. Schick, L.; Winthereik, B.R. Innovating Relations–or Why Smart Grid is not too Complex for the Public. *Sci. Technol. Stud.* **2013**, *26*, 82–102.
11. Kobus, C.B.; Klaassen, E.A.M.; Mugge, R.; Schoormans, J.P.L. A real-life assessment on the effect of smart appliances for shifting households' electricity demand. *Appl. Energy* **2015**, *147*, 335–343. [CrossRef]
12. Naus, J.; van der Horst, H.M. Accomplishing information and change in a smart grid pilot: Linking domestic practices with policy interventions. *Environ. Plan. C Politics Space* **2017**, *35*, 379–396. [CrossRef]
13. Powells, G.; Bulkeley, H.; Bell, S.; Judson, E. Peak electricity demand and the flexibility of everyday life. *Geoforum* **2014**, *55*, 43–52. [CrossRef]
14. Smale, R.; van Vliet, B.; Spaargaren, G. When social practices meet smart grids: Flexibility, grid management, and domestic consumption in The Netherlands. *Energy Res. Soc. Sci.* **2017**, *34*, 132–140. [CrossRef]
15. Nyborg, S.; Røpke, I. Constructing Users in the Smart Grid—Insights from the Danish eFlex Project. *Energy Effic.* **2013**, *6*, 655–670. [CrossRef]
16. Kessels, K.; Kraan, C.; Karg, L.; Maggiore, S.; Valkering, P.; Laes, E. Fostering residential demand response through dynamic pricing schemes: A behavioural review of smart grid pilots in Europe. *Sustainability* **2016**, *8*, 929. [CrossRef]

17. Skjølsvold, T.M.; Jørgensen, S.; Ryghaug, M. Users, design and the role of feedback technologies in the Norwegian energy transition: An empirical study and some radical challenges. *Energy Res. Soc. Sci.* **2017**, *25*, 1–8. [CrossRef]

18. Fell, M.J.; Shipworth, D.; Huebner, G.M.; Elwell, C.A. Public acceptability of domestic demand-side response in Great Britain: The role of automation and direct load control. *Energy Res. Soc. Sci.* **2015**, *9*, 72–84. [CrossRef]

19. Hansen, M.; Hauge, B. Scripting, control, and privacy in domestic smart grid technologies: Insights from a Danish pilot study. *Energy Res. Soc. Sci.* **2017**, *25*, 112–123. [CrossRef]

20. Kendel, A.; Lazaric, N.; Maréchal, K. What do people 'learn by looking'at direct feedback on their energy consumption? Results of a field study in Southern France. *Energy Policy* **2017**, *108*, 593–605. [CrossRef]

21. Bulkeley, H.; Powells, G.; Bell, S. Smart grids and the constitution of solar electricity conduct. *Environ. Plan. A* **2016**, *48*, 7–23. [CrossRef]

22. van Mierlo, B. People's Involvement in Residential PV and their Experiences. In *Photovoltaic Solar Energy: From Fundamentals to Applications*; Reinders, A., Verlinden, P., Freundlich, A., Eds.; John Wiley & Sons, Ltd.: Hoboken, NJ, USA, 2017; pp. 634–645.

23. Hier Opgewekt. *Lokale Energiemonitor 2017 [Local Energy Monitor 2017]*; Hier Opgewekt: Utrecht, The Netherlands, 2017.

24. Arentsen, M.; Bellekom, S. Power to the people: Local energy initiatives as seedbeds of innovation? *Energy Sustain. Soc.* **2014**, *4*, 1–12. [CrossRef]

25. Blanchet, T. Struggle over energy transition in Berlin: How do grassroots initiatives affect local energy policy-making? *Energy Policy* **2015**, *78*, 246–254. [CrossRef]

applied sciences

MDPI

Article

Transaction Mechanism Based on Two-Dimensional Energy and Reliability Pricing for Energy Prosumers

Eunsung Oh [1] and Sung-Yong Son [2],*

[1] Department of Electrical and Electronic Engineering, Hanseo University, Chungcheongnam-do 31962, Korea; esoh@hanseo.ac.kr

[2] Department of Electrical Engineering, Gachon University, Gyeonggi-do 13120, Korea

*   Correspondence: xtra@gachon.ac.kr; Tel.: +82-31-750-5347

Received: 15 February 2019; Accepted: 26 March 2019; Published: 30 March 2019

Abstract: Prosumers, users that consume and produce energy, increase the diversity of energy system operations as distributed sources. However, they can reduce energy system reliability by increasing uncertainty. This study presents a novel transaction mechanism based on dynamic pricing for enhancing energy system reliability. The proposed dynamic energy-reliability pricing-based transaction mechanism (ERT) increases the controllability of prosumer uncertainties by a two-dimensional pricing mechanism based on time and reliability status, as compared to conventional time-based one-dimensional energy prices. Under the proposed ERT, utilities and prosumers exchange information about the utility price and the prosumer's intent in order to ensure that demand is met. A two-way information infrastructure built for prosumer energy trading is used for this task. The utility enhances system reliability using this information, and the prosumer increases revenue through pricing selection. The practical implementation of the proposed ERT is described for both utilities and prosumers. A case study using practical renewable generation data revealed that the proposed ERT improves not only the reliability factor of the utility but also prosumer revenue as compared to a conventional energy-based dynamic pricing case. It is also shown that an economical optimum point that maximizes prosumer net revenue exists when electrical energy storage (EES) is applied to enhance performance. Increasing to the EES capacity provided room for uncertainty management, net revenue is improved, but the economic burden by the EES cost is increased. Under the proposed ERT, the optimal point results in greater EES capacity and higher net revenue enhancement than the conventional case.

Keywords: dynamic pricing; electricity pricing; prosumer; reliability; renewable energy; uncertainty

1. Introduction

Along with the development of the smart grid, more and more distributed generation has been deployed in power systems. This often includes renewable energy sources due to their availability, applicability, and environmentally friendly nature [1]. With distributed renewable generation, traditional passive consumers, such as households and small businesses, are now participating as producers in the electricity market [2]. These newly introduced users are called prosumers, as they both share surplus energy generated domestically as well as consume energy from the energy utility [3].

The presence of prosumers is expected to bring some benefits. These include: (1) a modular and tailored energy service to specific end uses, which can vary in quality; (2) grid diversification, by encouraging the large amount of distributed generation deployed to complement decentralized storage options; and (3) efficient wholesale market operations, with minimal transmission congestion and constraints [4]. However, an increase in renewable sources and decentralization can cause reliability problems, introducing multiple unknowns and risks that must be addressed and managed in grid operation [5].

In order to address the reliability problem, a direct approach based on control, or a soft or indirect approach based on pricing, can be used. The direct approach is hardware-centric. To solve the grid operation issues, additional equipment, such as smart inverters [6], static synchronous compensators (STACOM) [7], and/or electric energy storage (EES) [8,9], can be deployed. The grid operation structures are improved to enable cooperation with the distributed resources via the introduction of advanced distribution network operating systems [10]. Applying this direct approach is effective in terms of operations. However, there are additional investment and regulatory problems. Alternatively, the soft/indirect approach pursues a similar operation effect statistically by providing additional signals to participants. This approach requires more sophisticated technologies, and it lessens the direct investment of utilities necessary to operate the grid.

A pricing-based approach is a typical mechanism for indirect control [11]. It is a demand-side management mechanism that improves grid reliability by indirectly controlling the prosumer's energy usage to the price factor, such as demand charges and energy charges [12]. However, conventional pricing that consists of demand charges and energy charges usually has limitations in considering load or generation uncertainties [13]. Under conventional pricing systems, distributed renewable generators have limited responsibility and limited rewards for their uncertainties, as the pricing is determined regardless of generation uncertainty [14].

Dynamic pricing, such as real-time pricing (RTP) [15] or time-of-use (TOU) pricing [16], can lead generators to produce electricity at higher-price times. However, this method cannot help to decrease generation uncertainty. From the viewpoint of utilities that operate grids, generation uncertainty is important in operational planning. Here, uncertainty can be represented as generation predictability. As such, a more advanced pricing method has been introduced to improve generation predictability by assigning the responsibility of prediction to generators and by instituting a penalty and reward system for their commitments. Even in this advanced pricing system, utilities have limited information for generation prediction, because even though prosumers commit their generation amount, they do not have information about the uncertainty of their generation amount. Prosumers with renewable generation capabilities may try to keep their commitments to avoid or reduce penalties [17]. The utilities, however, are unaware of the situation and intend to honor prosumers' generation commitments, and thus can only estimate the amount of the generation based on historical statistical information. The exact same issue occurs in load estimation. It means that an information-driven transaction is required in active prosumer environments [18].

In this study, a novel transaction mechanism based on dynamic pricing is proposed to pursue improved utility grid operation. The basic concept of the proposed mechanism increases each benefit of utility and prosumers by exchanging more information. For this purpose, it is described as the proposed mechanism and the operation sequence. As a first step, the utilities can announce multiple uncertainty pricing bands beforehand that may vary depending on time, and the prosumers can select a reliability band appropriate for their situation. In the proposed mechanism, prosumers commit both the amount of generation along with the uncertainty of that amount. Thus, utilities have access to more detailed information about generation uncertainties than under the current system. A case study shows that, under the proposed mechanism, prosumers achieve more revenue and grid reliability is improved for utility compared to the conventional pricing case. This is because both utilities and prosumers have an opportunity to maximize their benefits via additional information about uncertainty. In addition, by performance characteristic and economic sensitivity analysis, it is discussed that the proposed mechanism is adaptively worked and has an economically optimal point related to resource characteristics. This mechanism is described from the generation point of view, but it works in the same way for demand.

The rest of this paper is organized as follows. Section 2 describes the proposed dynamic-pricing-based transaction mechanism and compares it to conventional energy-based dynamic pricing. Section 3 discusses how to implement the proposed mechanism in energy systems. Section 4 demonstrates a measurement study using real wind generation data applied to the proposed dynamic pricing mechanism. Section 5 concludes the paper.

2. Dynamic Energy-Reliability Pricing-Based Transaction Mechanism

2.1. Energy-Reliability Pricing-Based Transaction Mechanism

A conventional energy-based dynamic price is announced by a utility or retailer, and consumers can control their demand considering the publicly announced price. The utility provides price information in one-way communication, and a consumer responds with their energy usage for each time period *t*. Here, the information is the only price on time, so the information can be considered one-dimensional. The conventional pricing is expressed as

$$P_{Conv} = \left\{ p_1^C, \cdots, p_t^C, \cdots, p_T^C \right\}, \tag{1}$$

where *T* is the operation time period.

Under the proposed dynamic energy-reliability pricing-based transaction mechanism (ERT), however, the utility announces reliability bands on time as well as price on time to prosumers, as shown in Figure 1. A prosumer responds to the utility with their expected demand and selected reliability band. The utility can announce the reliability price for each band for each time. Therefore, the communication is two-way. The proposed ERT is two-dimensional, since both the energy price and reliability price are included, as follows:

$$P_{ERT} = \left\{ \alpha_1 p_1^E, \cdots, \alpha_t p_t^E, \cdots, \alpha_T p_T^E \right\}, \tag{2}$$

where α_t is a relative price in the range of [0, 1], as shown in Figure 1. The relative price is a multivariable function according to reliability $\gamma_t \in (0, 1]$ and band selection indicator $\beta_{i,t} \in \{0, 1\}$,

$$\alpha_t(\beta_{i,t}, \gamma_t) = \sum_{i \in B} \beta_{i,t} f_i(\gamma_t), \tag{3}$$

where *B* is a band set, e.g., three band (Band 1, Band 2, Band 3) in Figure 1.

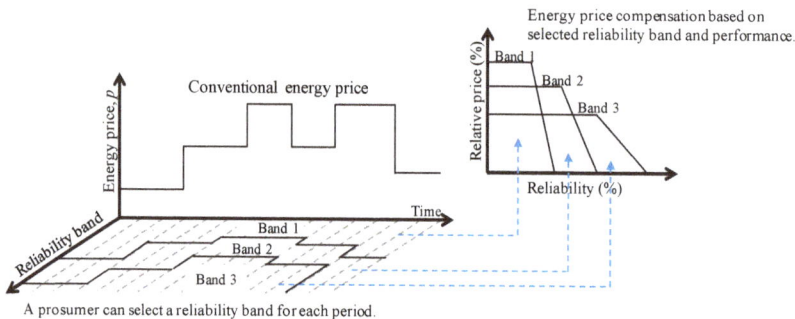

Figure 1. The proposed dynamic energy reliability pricing concept.

A comparison between the conventional pricing mechanism and the proposed ERT is summarized in Table 1.

Energy usage intentions are indirectly controlled in conventional price-based systems using unit prices for specific times. Under the ERT, additional information is used to understand prosumers' intention regarding energy usage, and uncertainty is managed differently than in conventional price-based systems.

Table 1. Comparison between the conventional pricing mechanism and the proposed dynamic energy-reliability pricing-based transaction mechanism (ERT).

	Communication	Pricing
Conventional pricing	One-way	One-dimensional (Time)
Proposed ERT	Two-way	Two-dimensional (Time and reliability)

2.2. ERT Operation Procedure

Figure 2 shows the operation sequence of the proposed ERT mechanism.

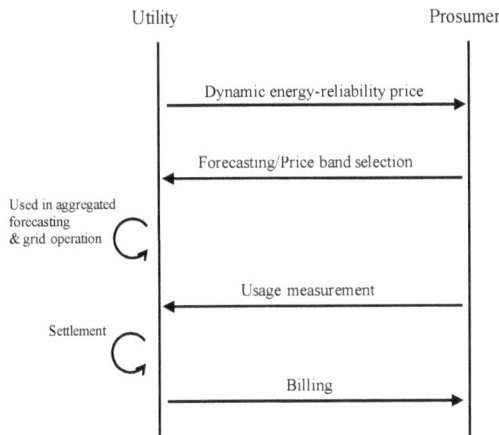

Figure 2. The ERT operations sequence.

(Step 1. ERT announcement) First, a utility or retailer announces the ERT, P_{ERT}, which consists of energy prices and corresponding reliability bands. Basically, the energy price would similarly change over time but should be differently rewarded depending on the quality of a prosumer's resources.

(Step 2. Reliability forecasting and band selection) Under the ERT, prosumers should forecast their energy consumption or production, $\hat{E} = \{\hat{e}_1, \cdots, \hat{e}_t, \cdots \hat{e}_T\}$; this is an essential contribution from the prosumer. In addition, they need to forecast the uncertainty of their energy forecast $\hat{\gamma} = \{\hat{\gamma}_1, \cdots, \hat{\gamma}_t, \cdots \hat{\gamma}_T\}$. When a prosumer understands their own resource characteristics, they can select the most appropriate reliability band or strategically change their operation plans to maximize their expected benefit by deducing potential risks, as follows:

$$\beta_{i^*,t} = \underset{i \in B}{\arg\max} \alpha_t(\beta_{i,t}, \hat{\gamma}_t) p_t \hat{e}_t, \quad \forall t \in T. \tag{4}$$

Prosumers can even decide to invest in additional infrastructures, such as EES, to increase their forecast accuracy and to reduce their uncertainty. Once a prosumer forecasts their energy usage and uncertainty considering their resource characteristics and operation plan, the forecast and selected uncertainty level-based band are sent to the utility.

(Step 3. Settlement and billing) The utility aggregates the individual prosumer forecasts and commitments, incorporating this information into its own forecasts and grid operation plans. The basic assumption here is that, by obtaining prosumers' forecast and band selection information, a utility can increase its own forecasting accuracy regarding both amount and uncertainty. Even if the utility's forecast based on this information includes errors, it can minimize its financial risk by penalizing prosumers for inaccurate forecasting. Based on a prosumers' band price selection, the utility or prosumer can perform settlement based on the prosumer's energy usage. Then, billing occurs as usual,

$$\text{Energy bill} = \sum_{t \in T} \alpha_t(\beta_{i^*, t}, \gamma_t) p_t e_t. \tag{5}$$

3. ERT Implementation Issues

To apply the ERT, various methodologies can be considered for a dynamic energy-reliability price on the utility side and forecasting and band selection on the prosumer side. It is structured according to the operating purpose of the system and the resource characteristics that the players (utility and prosumer) have. This section presents ERT implementation issues from both the utility and prosumer sides and provides basic guidance for applying the ERT.

3.1. Utility Side

When applying the proposed ERT, the grid utility announces not only the energy price but also the reliability price and band for each operational time horizon. The energy price is determined by cost-balancing of the utility and prosumer, as under the existing pricing concept. The reliability price and band are decided considering utility requirements and prosumer characteristics.

3.1.1. Energy Price Design

The basic principle involved in a price is to achieve a balance between the utility's income from grid operation and the prosumers' benefit from supplying/consuming energy. The price at each time is designed considering business sustainability and economic efficiency [19]. The proposed pricing mechanism requires more information about uncertainty generated from the prosumer side. This information enhances grid reliability and reduces the operational costs associated with the risk of uncertainty. In this manner, the utility sets up a price at each time considering the grid reliability enhancement and business sustainability. Figure 3 shows a TOU pricing example of Korea Electric Power Company (KEPCO) [20]. It is designed to reflect the temporal energy demand and supply change. The value of conventional pricing can be used as a baseline control,

$$P_t^E = p_t^C, \quad \forall t \in T. \tag{6}$$

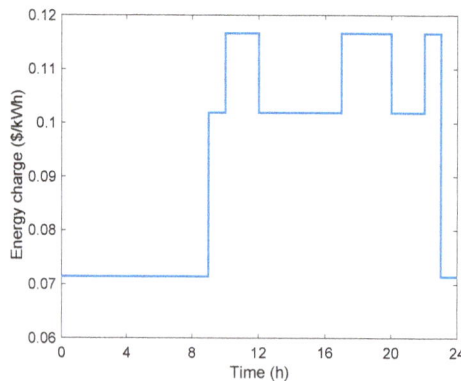

Figure 3. A time-of-use (TOU) pricing example of Korea Electric Power Company (KEPCO).

3.1.2. Reliability Band and Price Design

In the proposed ERT, the reliability band and price are adjusted according to the utility's requirements. Basically, the reliability band is narrow and the price is set high where the utility requires high reliability. On the other hand, however, this leads to discrimination between prosumers with different characteristics.

Another way to design the reliability band and price is to consider prosumer priority in terms of the utility. Figure 4 shows the contribution of wind generation at net demand from the Bonneville Power Administration (BPA) balancing authority in 2016 [21]. Relative to contribution, the utility decides the prosumer priority, and it is used as a reference to determine the reliability band range. In Figure 4, the average contribution of wind generation is around 22%, so the priority ω_t is set as 3 for 25% or more, 2 for 25–20%, and 1 for less than 20% as an example. Using the priority, the utility designs the relative price function in (3). Narrower reliability bands will be allocated to the higher priorities to effectively manage uncertainty, as follows:

$$f_i(\gamma_t) = \begin{cases} 1, & \gamma_t \le \frac{1}{\omega_t}, \\ -\gamma_t + 1, & otherwise. \end{cases} \tag{7}$$

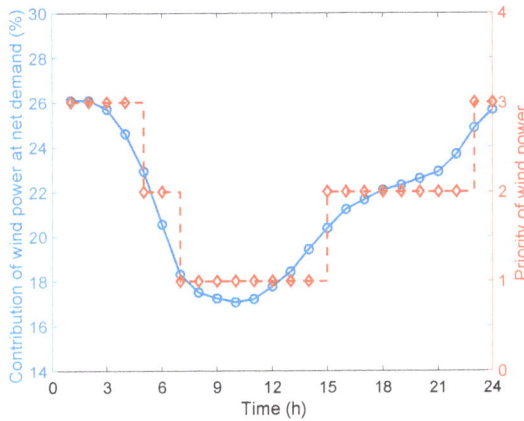

Figure 4. Contribution of wind generation at net demand. The priority is set as 3 for 25% or more, 2 for 25–20%, and 1 for less than 20% related to the contribution as an example.

3.2. Prosumer Side

Under a conventional price, prosumers passively receive an energy bill related to their usage, whereas, under the ERT, their energy bill is dynamically determined through their own operation and selection. To maximize their benefits under the ERT, a prosumer estimates their own demand and selects their own reliability band.

3.2.1. Demand Forecasting

The ERT suggests a different price on time and reliability status. To decide the best way to reduce an energy bill and to enhance their benefits, a prosumer should know their own situation. The basic way to estimate demand is using a conventional load forecasting method, such as the customer baseline load (CBL) [22,23] or the renewable generation forecasting method [24,25]. Another way is controlling demand using prosumer characteristics. Internally controlling the load and generation, a prosumer selects a demand to maximize their benefit. However, this requires additional operation with high risk and complexity caused by two uncertain demand sets.

3.2.2. Reliability Band Selection

To select a reliability band, the reliability of the forecasted value at each operational time horizon is required. Conventional forecasting methods, such as CBL, do not provide the value nor provide it as a probabilistic average. The reliability of the forecasted value is related to the accuracy of the demand forecasting. Figure 5 shows the relationship between the gradient of forecasting and the accuracy of BPA's wind generation [26,27]. Prosumers can estimate the reliability as an example method,

$$\hat{\gamma}_t = \kappa |\hat{e}_t - \hat{e}_{t-1}|, \tag{8}$$

and, substituting (8) into (4), they select the reliability band.

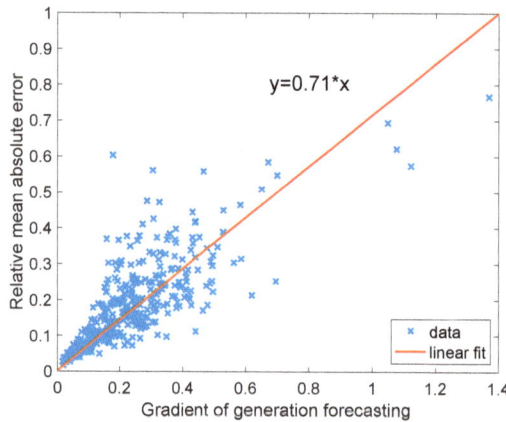

Figure 5. The relationship between the gradient of generation forecasting and the relative mean absolute error of Bonneville Power Administration (BPA)'s wind generation.

Similar to the internal control for demand forecasting, the reliability is also enhanced using additional control afforded by energy storage. The selection of a high-reliability band with additional control enhances potential benefits. However, this increases both the operational burden of prosumers and their costs from additional equipment (such as energy storage).

4. Case Study

In this section, the performance of the proposed ERT is evaluated compared to a conventional penalty-based dynamic pricing mechanism, and the characteristics of the proposed method are discussed.

4.1. Experimental Environment

A case study was performed where the proposed ERT was applied as the energy price when the prosumers supply the energy. Although it only focuses on the energy supply price for the prosumer as a producer, the proposed work can also be applied to the case when the prosumer purchases energy as well.

Figure 6 shows the incentive and penalty plan of conventional pricing and the ERT. The incentive and penalty band of the proposed price has a different range and slop at each band shown as lines in Figure 6a. By adjusting these bands, the utility controls the prosumer's responsiveness. As shown in Figure 6b, the dynamic energy-reliability pricing is determined by combining the incentive-penalty band to control the reliability and the energy price on time to manage energy consumption.

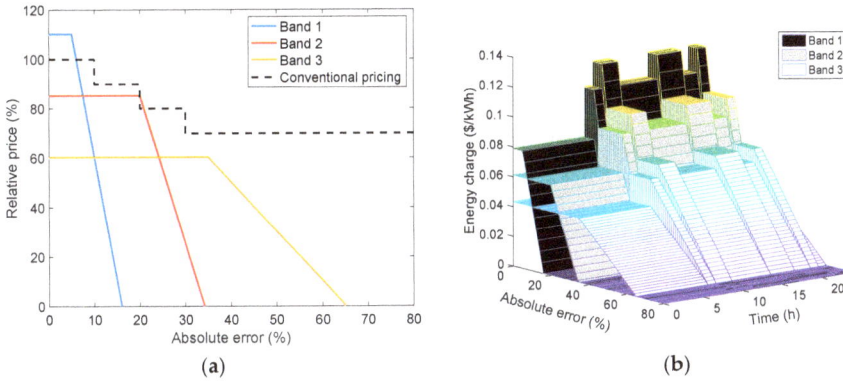

Figure 6. Incentive and penalty plan of a conventional pricing mechanism and the ERT. (a) Incentive-penalty plan; (b) Energy-reliability pricing.

To compare the performance, an energy price with a step-down penalty charge $\alpha_t^c(\gamma_t)$ [28] was considered as the conventional case, expressed as conventional pricing shown as a dashed line in Figure 6a. For a fair comparison, the band of the proposed price, $\alpha_t(\beta_{i,t}, \gamma_t)$, is designed to have the same revenue as the conventional pricing case when they do not make any effort,

$$\sum_{t \in T} \alpha_t^c(\gamma_t) p_t e_t = \sum_{t \in T} \alpha_t(\beta_{i,t}, \gamma_t) p_t e_t. \tag{9}$$

Assuming a three-band case, the values are numerically determined in order to have the same expected revenue using the probability distribution of absolute errors as shown in Figure 7.

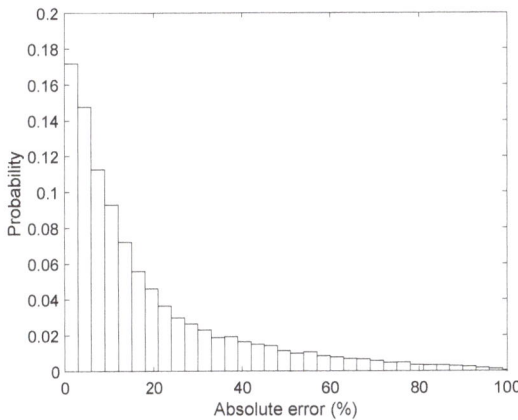

Figure 7. Probability distribution of absolute error for generation forecasting.

Under both the conventional pricing case and the proposed ERT, the prosumer (producer in this case) achieves the same annual revenue when operating the generator without management response to the energy price. The value without management was used as the reference value, and, for confirming the performance, the value was evaluated when EES was added and optimally operated to maximize prosumer revenue.

The energy supply data used comes from the BPA, United States Department of Energy [21]. The BPA controls 42 wind plant sites concentrated in the Columbia River Gorge that spans northern

Oregon and southern Washington, and its total capacity in 2016 was around 4600 MW. Wind power generation data and forecasts over 360 days in 2016 were used and are presented in Figures 4, 5 and 7.

4.2. Performance Evaluation

For a transaction mechanism to work well, the benefits for both the prosumer and the utility should be preserved. The revenue for the prosumer and the reliability for the utility were considered to verify the performance of the proposed mechanism.

4.2.1. Prosumer Revenue

Figure 8 shows the revenue enhancement by the conventional pricing and the proposed ERT when the size of EES is increased to 30% of the installed generation capacity. The EES is operated to maximize prosumer revenue by reducing the error between the generated forecasting and the actual generation, so the revenue is enhanced in the dynamic prices. The revenue converges on a 4% enhancement over the conventional pricing, but the value in the proposed ERT grows continuously, up to a value three times higher than that under the conventional pricing. This shows that the proposed ERT is designed to work well in relation to prosumer dynamics, giving the prosumer greater motivation.

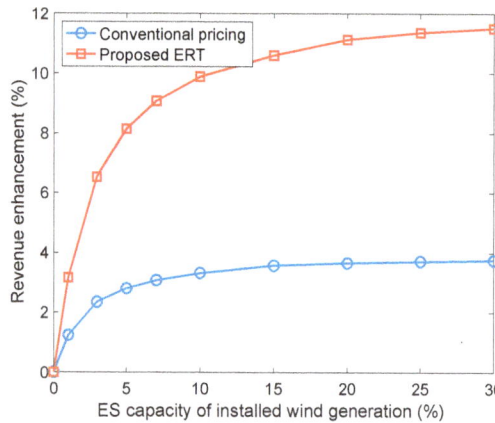

Figure 8. Comparison of user revenue enhancement under the conventional pricing mechanism and the proposed ERT.

4.2.2. Grid Reliability

For utility, grid reliability can be defined as such factors as voltage stability and frequency regulation state maintenance [29]. It is required for better situational awareness and operator assistance to improve grid reliability [5]. The predictability is an important term that affects reliability. In this work, the mean absolute error (MAE) between the actual generation and its forecasting is used as a performance metric to measure predictability. In Figure 8, the MAE improves with increasing EES capacity. It is said that grid reliability improves with increasing EES capacity. Additionally, the MAE under the proposed ERT has a lower value than that under the conventional pricing mechanism. This means that the implementation of the proposed ERT effectively improves grid reliability versus the conventional pricing case. However, the MAE is not monotonically improved, as shown by the proposed ERT case with 15% EES capacity of installed generation (the point in the dashed circle in Figure 9). This is because the performance of MAE not only relates to the price but also the generation characteristics.

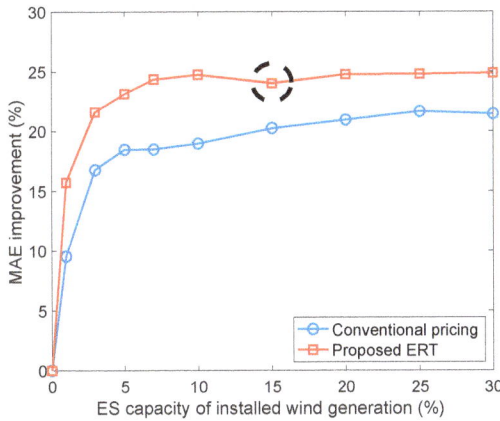

Figure 9. Comparison of grid mean absolute error improvement under the conventional pricing mechanism and the proposed ERT.

4.3. Characteristics

In the performance evaluation section, it is shown that the prosumer benefit and grid reliability with the proposed ERT is enhanced more than under the conventional pricing case. Furthermore, it is mentioned that the performance enhancement is related to the characteristics of the price and the resource. To explain that, it is shown how to change the band hit ratio under the conventional pricing case and the proposed ERT for various EES capacities. Band hit ratio means the normalized absolute error distribution at each band range as shown in Figure 6, after the EES operation to maximize the prosumer revenue under each price condition. The EES enhances the system's ability to achieve the operational objective, i.e., the prosumer revenue maximization in this work. Therefore, the results of increasing EES capacities show the direction of achieving the purpose of the system.

In Figure 10a, as the operational room increases, such as EES capacity, the second band hit ratio grows under the conventional price. This is because the average MAE without EES is about 18%, included in the second band of the conventional price. The results show that, under the conventional price, the prosumer works to maximize their benefit by reducing their average penalty. This is because the incentive-penalty is settled according to the result of operation under the conventional pricing case. However, under the ERT, the first band hit ratio and the out-of-band ratio are increased in relation to the EES capacity in Figure 10b. This means that the prosumer in the proposed ERT achieves enhanced benefits by maximizing the incentive. In the proposed ERT, the incentive-penalty is determined according to the previously selected band and operation result. Therefore, the prosumer operates the system considering the reliability as well as the benefit, as shown in Figure 10b. From these results, it can be said that the prosumer under the conventional pricing case operates the system in the form of an average maximization criterion, and a max-max-type operation is worked by the proposed ERT.

In addition, the out-of-band ratio increases from 10% EES capacity (the sixth bar in Figure 10b) to 15% EES capacity (i.e., the seventh bar in Figure 10b is larger than the others). This is why the MAE improvement under the proposed ERT is slightly reduced as shown in Figure 9. These results show that the proposed ERT effectively works in relation to the resource characteristics and gives the prosumer more motivation to control their resources than under the conventional pricing case.

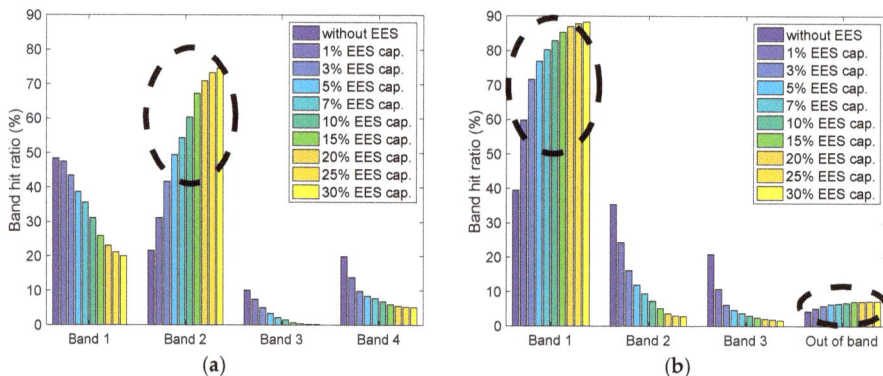

Figure 10. The change in band hit ratio under the conventional pricing mechanism and the proposed ERT with varying EES capacity. (**a**) Conventional price case; (**b**) Proposed ERT.

4.4. Economic Sensitivity Analysis

As shown in Figures 8 and 9, the performance under the dynamic price is enhanced by increasing the EES capacity. The EES is a good solution to manage uncertainty, but increasing the EES capacity reduces the revenue by growing the EES cost. In this section, we evaluated the net revenue considering the EES cost.

The net revenue is calculated as:

$$Net\ revenue = PW(\chi) \times Annual\ revenue - EES\ cost, \tag{10}$$

where $PW(\chi)$ is the current worth factor that fairly adjusts the revenue to accommodate future timeframes into the present value. Assuming the service time $\chi = 10$ (year) of EESs, a price escalation rate $r_e = 2.5$ (%/year), and a discount rate $r_d = 10$ (%/year), the worth factor is calculated as [30]:

$$PW(\chi) = \sum_{i=1}^{\chi} \frac{(1 + r_e)^{i-0.5}}{(1 + r_d)^{i-0.5}}. \tag{11}$$

The EES cost is constructed from the power-subsystem capacity cost and the energy-subsystem capacity cost. Based on the corresponding capacity costs of Li-ion batteries, it was estimated as $240/kW and $360/kWh, respectively [27].

Figure 11 shows the net revenue per year. Under the proposed ERT case illustrated as a solid line with square markers, the net revenue is greater than that without EES over the entire range, and the maximum net revenue enhancement achieved is about 6.6% when the EES capacity applied is 7% of installed generation. However, under the conventional price presented as a solid line with circle markers, the net revenue is greater than that without EES only when the EES capacity is less than 10% of the installed generation, and the maximum net revenue enhancement is about 1%, with 2% EES capacity of installed generation. As such, the proposed ERT is economically efficient and adaptively manages the uncertainty, and the conventional pricing is inefficient.

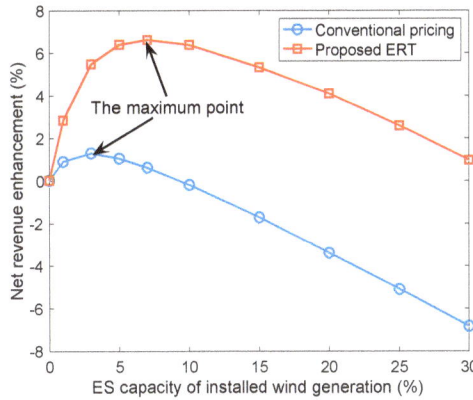

Figure 11. Net revenue under the conventional pricing mechanism and the proposed ERT.

4.5. Other Reliability Band Design Case

The performance of the proposed ERT is highly related to the reliability band design values. The value is decided by the utility, considering how best to manage prosumer behavior. Therefore, many kinds of values could be determined.

Figure 12 shows a comparison of the revenue enhancement by the conventional pricing (Case 0) and two ERTs (Case 1 and 2) when 10% EES capacity of installed generation is used. Case 1 is the ERT discussed in previous sections, and Case 2 has the design values expressed in Table 2. Each of the 360 points in Figure 12 expresses the daily prosumer revenue enhancements.

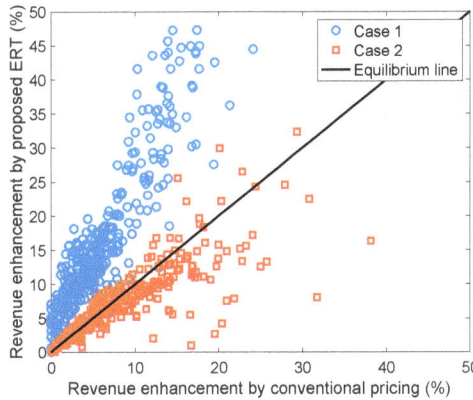

Figure 12. Comparison of revenue enhancement under the convention pricing mechanism and proposed ERTs.

Table 2. Price design values.

	Band 1		Band 2		Band 3		Band 4	
	Price	**Range**	**Price**	**Range**	**Price**	**Range**	**Price**	**Range**
Case 0	100	0–10	90	10–20	80	20–30	70	30-
Case 1	110	0–5	85	0–20	60	0–35	-	-
Case 2	100	0–10	85	0–40	70	0–70	-	-
	Case 0 expresses the conventional pricing case.							

In Case 1, the prosumer realizes more revenue than under the conventional price every day. The average prosumer revenue improvement has the same value under both the conventional pricing and the ERT in Case 2, but the prosumer revenues under the ERT in Case 2 have a higher or lower value than under the conventional price. This is because the proposed ERT varies according to the daily generation characteristics. However, in Case 2, the utility has more MAE improvement, 25.1%, versus 19.0% for the conventional system, as presented in Table 3. In addition, the MAE change also shows a similar pattern as presented in Figure 10b. These results verify that the proposed ERT is an effective way to manage grid reliability without loss of prosumer benefits.

Table 3. Price performance (%).

	EES Capacity of Installed Generation						
	1%	**5%**	**10%**	**15%**	**20%**	**25%**	**30%**
	Revenue enhancement						
Case 0	1.2	2.8	3.3	3.6	3.7	3.7	3.7
Case 1	3.2	8.1	9.9	10.6	11.1	11.4	11.5
Case 2	1.2	2.9	3.3	3.5	3.6	3.6	3.6
	MAE improvement						
Case 0	9.5	18.5	19.0	20.2	20.9	21.6	21.6
Case 1	15.7	23.1	24.7	24.0	24.8	24.8	24.9
Case 2	15.2	23.7	25.1	25.9	26.2	27.6	27.8

Comparing the results of Case 1 and Case 2 in Table 3, the greater revenue enhancement case yields less MAE improvement. Revenue and MAE are the performance metrics of the prosumer and the utility, respectively. Thus, the trade-off between revenue enhancement and MAE improvement can be used as the band design strategy.

5. Discussion

The proposed ERT has both advantages and disadvantages compared to conventional price-based systems, as summarized in Table 4.

Table 4. Advantages and disadvantages of the ERT.

	Utility/Retailer	**Prosumer**
Advantages	Enhanced forecasting with additional information Promotes self-investment, deferring direct investment	Flexible investment and operation choices Increased opportunity to maximize benefits
Disadvantages	Increase in operational complexity Increased grid operation risks Increase in incentives	Increase in operational complexity Increased risk of penalty

From the viewpoint of the utility or retailer, by using additional information provided by prosumers, forecasting accuracy can be improved as shown in Table 3 compared to the conventional simple pricing case. Here, the risk caused by forecasting error can be shifted from the utility or retailer to prosumers. Furthermore, by providing an additional incentive to prosumers for enhanced energy usage and uncertainty forecasting, prosumer self-investment can be promoted, deferring direct investment by the utility or retailer. However, the system becomes more complex when an ERT is implemented, and additional settlement and billing solutions are required. The grid operation risks can increase when prosumers forecast intentionally wrongfully. The utility or retailer should provide more incentives to prosumers for the proposed mechanism even though they can evade direct investment for facilities.

From the prosumer viewpoint, the proposed mechanism provides more room in selecting the means to maximize benefits. For example, when a prosumer possesses a resource with high uncertainty, the prosumer can select a low-reliability price band to minimize forecasting error. Alternatively, they can invest in energy storage to decrease uncertainty, and can then select a high-reliability price band to maximize profit as shown in Table 3. Conversely, the prosumer now has additional responsibility for forecasting and price selection, increasing system and operation complexity.

The more price differences between bands in the ERT there are, the stronger the motivation to the participating prosumers. As a result, the net revenue or profit of the prosumers and the reliability of the utility can both be increased. However, the utility would provide more incentives to prosumers. Therefore, the utility should be careful in ERT price design not to provide unnecessary incentives obtaining excessive reliability. ERT bands can be designed as the combination of the maximum price, the range of absolute error for the maximum price zone, and the shape of the declining slope. It means that various price models can be designed with the ERT.

6. Conclusions

This study investigated an ERT as a means to enhance energy system reliability. Compared to a conventional dynamic energy pricing on time only, the ERT is designed to provide a price in terms of reliability range as well as time. Using the two-way information exchange of price from the utility and prosumer situation and reliability band selection, both the utility and the prosumer are expected to have opportunities to maximize their benefits. The proposed mechanism can motivate prosumers to participate in the system reliability improvement by incentivizing or penalizing them depending on the contribution. A numerical case study using BPA data validates the effectiveness of the proposed ERT and shows that both the reliability of the utility and prosumer revenue is enhanced compared to a conventional energy-based dynamic pricing mechanism. The result shows that the prosumers' revenue enhancement is 6.6% when the EES capacity applied is 7% of installed wind generation. It is also shown that the revenue enhancement and reliability improvement are higher under the proposed ERT than under the conventional pricing case by applying two band ERT prices. In addition, the advantages and disadvantages of the proposed ERT are discussed. This suggests guidance in adopting the proposed ERT as a transaction mechanism to enhance the energy system's reliability. The study can be extended for appropriate or optimal price-band model design using the proposed ERT.

Author Contributions: S.-Y.S. conceptualized and designed the structure of the manuscript. E.O. performed data analysis and investigation. E.O. wrote the original draft, and S.-Y.S. reviewed and edited the manuscript. All the authors have read and approved the final manuscript.

Funding: The work of E.O. was supported by the National Research Foundation of Korea through the Korean Government (Ministry of Science and ICT) under Grant 2017R1E1A1A03070136. The work of S.-Y.S. was supported by the Korea Institute of Energy Technology Evaluation and Planning and the Ministry of Trade, Industry and Energy of the South Korea under Grant 20168530050080.

Conflicts of Interest: The authors declare no conflict of interest.

References

1. Hossain, M.S.; Madlool, N.A.; Rahim, N.A.; Selvaraj, J.; Pandey, A.K.; Khan, A.F. Role of smart grid in renewable energy: An overview. *Renew. Sustain. Energy Rev.* **2016**, *60*, 1168–1184. [CrossRef]
2. Cai, Y.; Huang, T.; Bompard, E.; Cao, Y.; Li, Y. Self-sustainable community of electricity prosumers in the emerging distribution system. *IEEE Trans. Smart Grid* **2017**, *8*, 2207–2216. [CrossRef]
3. Rathnayaka, A.J.D.; Potdar, V.M.; Dillon, T.S.; Hussain, O.K.; Chang, E. A methodology to find influential prosumers in prosumer community groups. *IEEE Trans. Ind. Inform.* **2014**, *10*, 706–713. [CrossRef]
4. Parag, Y.; Sovacool, B.K. Electricity market design for the prosumer era. *Nat. Energy* **2016**, *1*, 16032. [CrossRef]
5. Moslehi, K.; Kumar, R. A reliability perspective of the smart grid. *IEEE Trans. Smart Grid* **2010**, *1*, 57–64. [CrossRef]
6. Kamel, R.M. New inverter control for balancing standalone micro- grid phase voltages: A review on MG power quality improvement. *Renew. Sustain. Energy Rev.* **2016**, *63*, 520–532. [CrossRef]
7. Sirjani, R.; Jordehi, A.R. Optimal placement and sizing of distribution static compensator (D-STATCOM) in electric distribution networks: A review. *Renew. Sustain. Energy Rev.* **2017**, *77*, 688–694. [CrossRef]

8. Oh, E.; Son, S.-Y.; Hwang, H.; Park, J.-B.; Lee, K.Y. Impact of demand and price uncertainties on customer-side energy storage system operation with peak load limitation. *Electr. Power Compon. Syst.* **2015**, *43*, 1872–1881. [CrossRef]

9. Tavakkoli, M.; Pouresmaeil, E.; Godina, R.; Vechiu, I.; Catalao, J.P.S. Optimal Management of an Energy Storage Unit in a PV-Based Microgrid Integrating Uncertainty and Risk. *Appl. Sci.* **2019**, *9*, 169. [CrossRef]

10. Gungor, V.C.; Sahin, D.; Kocak, T.; Ergut, S.; Buccella, C.; Cecati, C.; Hancke, G.P. A survey on smart grid potential applications and communication requirements. *IEEE Trans. Ind. Inform.* **2013**, *9*, 28–42. [CrossRef]

11. Khan, A.R.; Mahmood, A.; Safdar, A.; Khan, Z.A.; Khan, N.A. Load forecasting, dynamic pricing and DSM in smart grid: A review. *Renew. Sustain. Energy Rev.* **2016**, *54*, 1311–1322. [CrossRef]

12. Bu, S.; Yu, F.R.; Liu, P.X. Dynamic pricing for demand-side management in the smart grid. In Proceedings of the 2011 IEEE Online Conference on Green Communications, New York, NY, USA, 26–29 September 2011.

13. Dutta, G.; Mitra, K. A literature review on dynamic pricing of electricity. *J. Oper. Res. Soc.* **2017**, *68*, 1131–1145. [CrossRef]

14. Simshauser, P. Distribution network prices and solar PV: Resolving rate instability and wealth transfers through demand tariffs. *Energy Econ.* **2016**, *54*, 108–122. [CrossRef]

15. Siddiqi, S.N.; Baughman, M.L. Reliability differentiated real-time pricing of electricity. *IEEE Trans. Power Syst.* **1993**, *8*, 548–554. [CrossRef]

16. Hamalainen, R.P.; Mantysaari, J.; Ruusunen, J.; Pineau, P.O. Cooperative consumers in a deregulated electricity market—Dynamic consumption strategies and price coordination. *Energy* **2000**, *25*, 857–875. [CrossRef]

17. Oh, E.; Son, S.-Y. Pair Matching Strategies for Prosumer Market Under Guaranteed Minimum Trading. *IEEE Access* **2018**, *6*, 40325–40333. [CrossRef]

18. da Silva, P.G.; Ilić, D.; Karnouskos, S. The impact of smart grid prosumer grouping on forecasting accuracy and its benefits for local electricity market trading. *IEEE Trans. Smart Grid* **2014**, *5*, 402–410. [CrossRef]

19. Reneses, J.; Gomez, T.; Rivier, J.; Angarita, J.L. Electricity tariff design for transition economies: Application to the Libyan power system. *Energy Econ.* **2011**, *33*, 33–43. [CrossRef]

20. *Electric Rate Table*; Korea Electric Power Company (KEPCO): Naju, Korea, November 2013.

21. Wind Generation and TOTAL load in the BPA Balancing Authority. Bonneville Power Administration, U.S. Dept. of Energy. Available online: https://transmission.bpa.gov/Business/Operations/Wind/ (accessed on 12 February 2019).

22. Mohajeryami, S.; Doostan, M.; Schwarz, P. The impact of customer baseline load (CBL) calculation methods on peak time rebate program offered to residential customers. *Electr. Power Syst. Res.* **2016**, *137*, 59–65. [CrossRef]

23. Mohajeryami, S.; Doostan, M.; Asadinejad, A.; Schwarz, P. Error analysis of customer baseline load (CBL) calculation methods for residential customers. *IEEE Trans. Ind. Appl.* **2017**, *53*, 5–14. [CrossRef]

24. Zhang, Y.; Wang, J.; Wang, X. Review on probabilistic forecasting of wind power generation. *Renew. Sustain. Energy Rev.* **2014**, *32*, 255–270. [CrossRef]

25. Chitsaz, H.; Amjady, N.; Zareipour, H. Wind power forecast using wavelet neural network trained by improved clonal selection algorithm. *Energy Convers. Manag.* **2015**, *89*, 588–598. [CrossRef]

26. Lange, M. On the uncertainty of wind power predictions—Analysis of the forecast accuracy and statistical distribution of errors. *J. Sol. Energy Eng.* **2005**, *127*, 177–184. [CrossRef]

27. Oh, E.; Son, S.-Y. Energy-storage system sizing and operation strategies based on discrete Fourier transform for reliable wind-power generation. *Renew. Energy* **2018**, *116*, 786–794. [CrossRef]

28. Alagappan, L.; Kahrl, F.; Bharvirkar, R. *Regulatory Dimensions to Renewable Energy Forecasting, Scheduling, and Balancing in India*; White Paper; National Association of Regulatory Utility Commissioners: Washington, DC, USA, March 2017.

29. *IEEE Guide for Electric Power Distribution Reliability Indices*; IEEE std. 1366-2012; IEEE Std: New York, NY, USA, May 2012.

30. Eyer, J.; Corey, G. *Energy Storage for the Electricity Grid: Benefits and Market Potential Assessment Guide*; Tech. Rep. SAND2010- 0815; Sandia National Laboratory: Albuquerque, NM, USA, February 2010.

applied
sciences

MDPI

Article

Publicly Verifiable Spatial and Temporal Aggregation Scheme Against Malicious Aggregator in Smart Grid

Lei Zhang [1] and Jing Zhang [2,1,*]

[1] College of Computer Science and Technology, Harbin Engineering University, Harbin 150001, China; lei_power@hrbeu.edu.cn
[2] School of Information Science and Engineering, Jinan University, Jinan 250022, China
* Correspondence: ise_zhangjing@ujn.edu.cn

Received: 16 November 2018; Accepted: 14 January 2019; Published: 31 January 2019

Abstract: We propose a privacy-preserving aggregation scheme under a malicious attacks model, in which the aggregator may forge householders' billing, or a neighborhood aggregation data, or collude with compromised smart meters to reveal object householders' fine-grained data. The scheme can generate spatially total consumption in a neighborhood at a timestamp and temporally a householder's billing in a series of timestamps. The proposed encryption scheme of imposing masking keys from pseudo-random function (PRF) between pairwise nodes on partitioned data ensures the confidentiality of individual fine-grained data, and fends off the power theft of n-2 smart meters at most (n is the group size of smart meters in a neighborhood). Compared with the afore-mentioned methods of public key encryption in most related literatures, the simple and lightweight combination of PRF with modular addition not only is customized to the specific needs of smart grid, but also facilitates any node's verification for local aggregation or global aggregation with low cost overhead. The publicly verifiable scenarios are very important for self-sufficient, remote places, which can only afford renewable energy and can manage its own energy price according to the energy consumption circumstance in a neighborhood.

Keywords: smart metering; spatial and temporal aggregation; privacy protection; internal attack; pseudo-random function

1. Introduction

With the development of Advanced Metering Infrastructure (AMI), Smart Metering as an important research subject in Smart Grid (SG) plays an increasingly important role and is closely associated with people's daily life [1,2]. Aggregating fine-grained metering data attracts householders and power suppliers. Power suppliers can calculate, forecast, and regulate accurately power distribution/price of the next period in real time while detecting fraud reports. Based on billing details and current power price, householders can adjust its appliance consumption module to reduce the power billing at the peak time; however, accessing householder's information on metering may cause security and privacy concerns, such as daily routines, the type of applications, etc. [1,2]. For this, in SG systems, one of the challenges faced by power big data is how to design one aggregation mechanism to balance the use of power data and individual privacy protection [2].

Protecting such sensitive private data from individual privacy threats needs to limit the authority of the utility company employee [2]. Namely, Supplier Billing System (SBS, sub-suppliers) will know only the total amount of the consumption for each customer, while the Energy Management System (EMS, demand prediction division) should know only the total consumption of customers in a certain region for each time period. To achieving the goals, smart metering systems often introduce the Meter Data Management System (MDMS), which stores the measured values of smart meters (SMs), and aggregates it before sending the aggregation to the SBS and EMS [2].

With the appearance of MDMS, another concern is upgrading, namely the malicious action of householders and regional MDMS employees. Unfortunately, a malicious householder may collude with the regional MDMS employee to report a false consumption to the SBS department; attackers may steal or forge power usage and consumption information. In addition, a regional MDMS employee may submit a fraudulent aggregation in a neighborhood. A World Bank report finds that each year over 6 billion dollars cannot post due to the energy theft and fraud report in the United States, in 2009, the FBI reported a wide and organized attempt that may have cost up to $400 million loss annually and power supplier suffered a great monetary loss [3]. To fend off this type of attack, it is desirable that suppliers or the public should detect the fraud profile from malicious aggregators or dishonest householders [4].

Privacy-preserving metering protocols have been discussed in lots of literatures [5–24]. They mainly focused on the studies of homomorphic aggregation [6–17,20,23,24], by which, aggregators can only obtain the fine-grained aggregation data within a certain region or householders' billing in a serial period while protecting individual privacy. However, most of them can only resist against single external or semi-trustable attack [11,13,14,19], and how to fend off internal attackers (e.g. aggregators or householders) is an open problem. Internal attackers can legally collect and store power consumption information of users; therefore, they pose a higher threat than external attackers [18].

Most of existing works [5–10,18,20,21,23,24] about additively homomorphic, multiplicative homomorphic, and their combination with other cryptography endeavored to address the problem. Most of them improved Paillier encryption [6–9] by their combination with other cryptography, such as stream ciphers [5,19,23] and modular addition [7,24], to prevent power suppliers/operators from intercepting individual user data, and to detect fraudulent from dishonest users. To ensure the integrity of transmitted messages and fend off attacks such as man-in the middle attack and denial-of-service attack against SG, signature and authentication methods are proposed in References [8,15–17].

Lu et al. [6] proposed a privacy-preserving, multi-dimensional metering aggregation scheme in a neighborhood-wide grid with piallier encryption, bilinear pairing and computational Diffie-Hellman (DH) methods. For resisting against internal attackers possessing private keys, Xiao [8] introduced a spatial and temporal aggregation and authentication scheme by randomizing Paillier encryption with Lagrange interpolation. Their protocol requires $O(n^2)$ bytes of inter-action between the individual meters as well as relatively expensive cryptography on the meters (public key encryption). Chen [9] also improved Paillier encryption and proposed a privacy-preserving aggregation scheme resisting at most t compromised servers in a control center with threshold protocol.

Dimitriou et al. [20] provided a verifiable publicly aggregation scheme against dishonest users that attempt to provide fraudulent data. Any user node in the community can prove its computation accuracy by zero-knowledge proof that the two encrypted message with different public keys corresponding to the same plaintext message. While we can prove our scheme costs lower overhead to resist fraudulent report from internal nodes.

Erkin et al. [23] adopted a stream cipher (e.g. RC4) to generate pseudo-random keys as masking keys between nodes to prevent internal nodes from possessing private keys. During the aggregation within a neighborhood, all masking random keys cancelled out and the aggregation value is revealed without compromising individual privacy based on the security properties of the Paillier encryption and stream cipher. We follow its Pseudo-Random Function and combine it with modular addition. The main difference from ours is they impose the random keys from PRF on the plaintext before encrypting it with Paillier cryptography, and send the encrypted message to all nodes. We set a security parameter k to represent the number of communicate nodes in a neighborhood and improve the encryption method by replacing the costly Paillier encryption with the simple and lightweight combination. More significantly, we supplement a publicly verifiable property to detect the fraudulent profile from malicious aggregators or dishonest user nodes.

Castelluccia et al. [19] protected individual data by imposing masking keys from RC4 on the plaintext data under the multi-level wireless sensors network model, However, the protection protocol cannot resist malicious aggregators, as the session keys are generated by the sink as the aggregator. We extend its PRF method into the peer-to-peer system model and propose a privacy-preserving scheme against maliciously internal attack.

In addition, traditional modular addition was adopted in [7,24] by partitioning individual plaintext data into n shares and exchanging them between nodes (n is the number size of users in a neighborhood). Flavio et al. [7] adopted Paillier encryption and modular addition, in which every user node partitions its meter reading into n shares and transmits the encrypted shares with different public keys to the aggregator, which aggregates the data with the same public key before sending the aggregation to the users. Finally, the aggregator collects the plaintext sums to obtain the final aggregation. The method is privacy-preserving; however, during each spatial aggregation, three message exchanges are required between every user and the aggregator. Thus, the number of homomorphic encryption per user increases linearly with n increases, and the communication overhead is $O(n^2)$ messages [20]. Jia et al. [24] also generated partitioned data with modular addition and imposed them on a high-order polynomial coefficient. The values of the polynomial at different points are transmitted to the aggregator which finds the coefficients of the polynomial with the private key and gains the aggregation, so the scheme is under the semi-trusted model and the aggregator is trustable. In addition, the computation overhead is relatively higher when k is increasing. As every node does the x^k polynomial operation before the matrix multiplication operation, the scheme increases greatly the computation overhead.

Ohara K et al. [4] summarized the function requirements during smart metering against internal attackers: calculating billing and obtaining statistics for energy management. We follow the statistic function requirements and the spatial and temporal scenarios in References [8,23] against malicious MDMS/aggregators or dishonest users:

(1) Spatial aggregation. A neighborhood-wide grid corresponds to a group of householders each equipped with a SG. They submit their encrypted meterings to the MDMS at a timestamp (e.g., 15 min). The latter aggregates homomorphically them before sending the aggregation to the EMS. During this aggregation, the individual data are confidential to the MDMS or the EMS.

(2) Temporal aggregation. A single SM submits its power consumption in a series of timestamps to the MDMS for the billing purpose. In this scenario, SBS charges the householders in serial timestamps.

Throughout this paper, we refer to the building area network (BAN) region as a neighborhood, and the regional MDMS as the regional gateway (GW), and the regional SBS as the control center (CC), respectively.

The main contribution can be summarized as follows:

(1) We design and implement a distributive, temporal and spatial aggregation scheme in the SG, in which every node sends and receives k encrypted message from k pairwise nodes distributively. The scheme provides spatial aggregation in a neighborhood at a fine-grained time scale (e.g. 15 min) and an individual temporal aggregation (e.g. monthly) in a series of timestamps for the billing purpose.

(2) The proposed encryption scheme minimizes the computation and communication overhead by replacing the costly public key cryptography adopted in most literatures with a combination of modular addition and PRF.

(3) The novel feature is that the masking keys are imposed on the partitioned data, and the latter are implemented by traditional modular addition. As the process of modular addition is processed by the node itself, other nodes cannot gain the true partitioned data, the masking key is only known to the pairwise nodes, and the combination ensures the confidentiality of individual data to any node including CC, aggregators, and n-2 nodes at most in a neighborhood.

(4) To detect malicious aggregators or dishonest users, we propose innovatively a publicly verifiable aggregation method. By this way, any user node in a neighborhood can receive the communication

flow, and verify the accuracy of local aggregation from other nodes or total aggregation from the aggregator without compromising individual fine-grained data.

(5)   The publicly available property for the aggregation also facilities householders regulating in time its current consumption module and consumption demand in the next time period, as by comparing their own consumptions with those of other nodes and checking if there is redundant power, householders can decide to store more energy or to sell excess power to the power supplier or other nodes. The scenarios are especially very important for self-sufficient, remote places, particularly, in developing countries, which can only afford renewable energy, such as wind turbines, solar panels, and carbon-based fuels [23].

The paper is organized as follows: in Section 2, we provide related preliminaries and formalize the system and attack models. In Section 3, we introduce our proposed aggregation scheme and correctness analysis. Security notions and proof are given in Section 4, followed by performance evaluation and comparison in Section 5. The conclusion is drawn in Section 6.

2. Preliminaries and Models

For ease of reading, we summarize the main notations in the paper in Table 1.

Table 1. Notations in the scheme.

Symbol	Meaning
HSM/SM	HAN smart meter/ user/user node/sm
N	The number of users in a BAN neighborhood1
k	The number of pairwise nodes for every user
K	Keystream based on stream cipher
M	RSA modular (large prime)
$x^i_{(j,d)}$	$user_i$ partitioned data into $user_j$ at timestamp d
x^i_d	$user_i$'s data at timestamp d
$r^i_{(j,d)}$	$user_i$'s pairwise key with $user_j$ at timestamp d
$E^i_{(j,d)}$	The encrypted form of $x^i_{(j,d)}$
sk_i	The secret key between CC and every node
$ind_i[s] (s = 1, \cdots, k)$	$user_i$'s pairwise nodes set in serial timestamps T
$LS(j,d)$	$user_j$'s locally spatial aggregation at timestamp d
$LT^i_{(j,d)} (j \in ind_i[s])$	$user_j$'s locally temporal aggregation for $user_i$ in T
$AT(i,T)$	$user_i$'s temporal aggregation in T
AS_d	Spatial aggregation in a neighborhood at timestamp d

2.1. Additively Homomorphic Encryption Based on The Keystream

Our security property partly comes from the stream cipher. The keystream generated from the pseudo-random function satisfies the security properties of the additively homomorphic encryption in the stream cipher. The basic idea [19] is denoted as follows:

Encryption is written as: $c = Enc_k(m + K) \bmod M$ where K is randomly generated keystream, m is the plaintext and $m, k \in [0, M - 1]$.

Decryption is described as: $Dec_k = c - K \bmod M$.

Additively homomorphic property of ciphertext are described as: $c_1 = Enc_{K1}(m_1)$ and $c_2 = Enc_{K2}(m_2)$; then, the aggregated ciphertext is expressed as: $c = c_1 + c_2 \bmod M = Enc_K(m_1 + m_2)$, where $K = K_1 + K_2 \bmod M$.

2.2. Pseudo-Random Keystream Generator—RC4

As a popular PRF generator, with secret keys between communication nodes, RC4 can generate a keystream. This secret key is pre-computed during the system initialization. As any stream cipher, the generated keystream can be used for encryption by combining it with the plaintext using bit-wise Exclusive-Or [19]. However our scheme is to replace the XOR (Exclusive-OR) operation typically found in stream ciphers with modular addition operation (+). To generate the keystream, RC4 needs two algorithms, i.e. Key-scheduling algorithm (KSA) and Pseudo-random generation algorithm (PRGA) [5,14].

KSA: KSA is to initialize a permutation with a variable length key between 40 and 2048 bits for PRGA.

PRGA: once the permutation initialization of KSA has been completed, the stream of bits is generated using the PRGA.

Algorithm 1: Key-scheduling algorithm (KSA)

Input: i = 0;
         j = 0 //Two 8-bit index-pointers
         S //The initial key keyed with a secret key
Output: S //A permutation of all 256 possible bytes
1. **for** (i = 0; i <= 255; ++ i)
2.   S[i] = i;
3. **end**
4. k = 0;
5. **for** (i = 0; i <= 255; ++ i)
6.   j = (j + s[i] + key[i mod keylength]) mod 256;
7.   k = S[i];
8.   S[i] = S[j];
9.   S[j] = k;
10. **end**

Algorithm 2: Pseudo-random generation algorithm (PRGA)

Input: i = 0;
         j = 0 //Two 8-bit index-pointers
Output: Z // Pseudo-random keystream
1. k = 0;
2. **for** (i = 0; i <= 255; ++ i)
3.   i = (I + 1) mod 256;
4.   j = (j + S[i]) mod 256;
5.   k = S[i];
6.   S[i] = S[j];
7.   S[j] = k;
8.   Z = S[(S[i] + S[j]) mod 256];
9. **end**

2.3. System Model

In our system model, we consider a typical SG communication architecture [8,9,11,15–17], as shown in Figure 1. It is based on the SG network model presented from the National Institute of Standards and Technology (NIST) and consists of six domains, i.e., the power plant, the transmission domain, the distribution domain and a CC, a residential GW, and the user domain. We mainly focus on how to report and aggregate the users' privacy-preserving data into the CC. Hence, the system model divides especially the BAN into numbers of Household area network (HAN) equipped with a SG and every BAN includes a GW and numbers of users.

CC: It acts as the SBS and EMS in reality. It needs to monitor the actual data on how much power is consumed at which timestamp in one BAN (neighborhood), how much power should be reserved for the next time period, and cumulative consumption for individual billing on a monthly basis, and how much power is being distributed to a specified neighborhood. In the paper, it is curious about the individual fine-grained data and may attempt to it as far as possible by all available resources, so it is assumed a semi-trusted entity.

GW: A powerful entity, acting as the local MDMS, represents a locality (e.g., a region within a building) is responsible for aggregating real-time spatial data in a neighborhood and individual temporal data in a series of timestamps, and then transmitting the aggregation to the CC. The employment of GW relieves CC of aggregation and reducing largely the communication latency. However, the cost that potentially malicious attacks done to users or power suppliers is unignorable, as discussed earlier. We assume it is a malicious entity here. A BAN GW represents a locality (e.g., a region within a neighborhood). For facilitating the communication between BAN GW and CC, WiMax and other broadband wireless technologies can be adopted. We consider a scenario that one BAN neighborhood covers a hundred or more HANs, so the longest distance from the BAN GW to a HAN is more than a hundred miles, so WiMax maybe more suitable for this kind of distance communication. Household Smart Meter (HSM): A bidirectional communication entity deployed at householders' premises. The modern SM is given a certain level of autonomy via trusted elements and the ability to collect, store, aggregate, and encrypt the usage data. Hence it has two interfaces—one interface is for reading power of householders and the other one acts as a communication GW. Even if we assume SM is tamper-resistant, it is not powerful as a GW, so it may be vulnerable to be compromised by the GW to infer the object users' data.

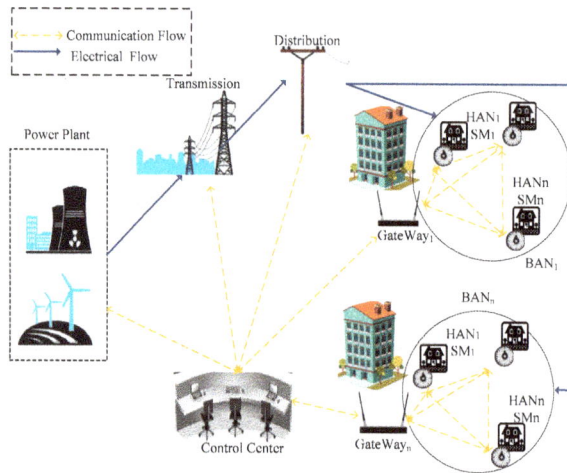

Figure 1. System model under consideration.

2.4. Communication Model

As can be seen in the Figure 1, all SMs connect each other in a neighborhood by WiFi technique, which constructs public verifiable foundation. Each user would select randomly k pairwise nodes in one round and can ensure that if $user_i$ chooses $user_j$, then $user_j$ chooses $user_i$ and the keys between them are opposite mutually. The value k as a security parameter can take any value from 2 to n, and depend on the specific application circumstance. The higher the value of k is, the higher the complexity is, and vice versa, and the scheme is more vulnerable to be attacked.

2.5. Data Model

Let x_d^i be the meter reading of the ith $(1 \leq i \leq N)$ user node at the dth $(1 \leq d \leq T)$ fine-grained timestamp, where N is the number of user in a BAN (a neighborhood-wide grid), and T is a billing period. At each fine-grained time index d, a neighborhood grid (over the entire BAN) spatially aggregated utility usage can be expressed as:

$$AS(d) = \sum_{i=1}^{n} x_d^i; \ d = 1, 2, \ldots, T \tag{1}$$

At the end of a billing period ($d = T$), a temporally aggregated utility usage data for the ith user is expressed as:

$$AT(i, T) = \sum_{d=1}^{T} x_d^i; \ i = 1 \text{ to } N \qquad (2)$$

2.6. Security Requirement and Attack Model

Within the system model, there are four types of actors involved in the meter data reporting process: the ith user (self), other users in the same neighborhood (BAN), the GW, and the CC. The CC requires the spatially aggregated fine-grained neighborhood usage data to optimize power delivery efficiency and the temporally aggregated user-specific utility usage data for the billing purpose. Hence, we stipulate the following security/privacy requirements:

Requirement **R1**. Fine-grained, individual utility data are private and should not be disclosed to CC, GW, or other users.

Requirement **R2**. Temporal aggregation for an individual user and spatial aggregation in one neighborhood cannot be tampered by the malicious aggregator or other internal nodes. For this, we envision a secure and reliable communication model comprising a verifiable publically method, which is customized to the correctness verification of the aggregation value of SG.

For this, our attack model is based on the malicious aggregator who attempts to tamper the aggregation value in a neighborhood and the billing value for individual users, or infers fine-grained meterings of the individual user by colluding with other n-2 compromised nodes at most. Following the above security requirements, different compositions of the attackers and actions may be grouped into the following attack types:

(1). External attack

External attackers may compromise the meterings of the object users by eavesdropping the communication flow between communication nodes through various eavesdropping malware.

(2). Malicious attack

False aggregation report. The aggregator may alter or drop maliciously any individual data, or tamper the aggregation data to the CC; any malicious user node may provide false local aggregation to the GW.

Collusion with compromised nodes. The aggregator may collude with compromised users to attempt to infer the uncompromised users' data.

(3). Semi-trustable internal attack

The curious CC or any user node can also acquire data through the public communication flow, such as the message from the user node to the GW or from the GW to the CC. They may infer the object user's fine-grained data by the public communication flow.

An attack is an arrangement that enables unauthorized parties to gain access to private data or to tamper secured data (even by the user itself) without being detected. In this work, we assume the SMs are tamper-resistant [7,20,23], and can perform the measurement and reporting operations normally, but do not exclude the possibility of tampering with local aggregation values by itself.

3. Proposed Scheme

3.1. Initialization Phase

3.1.1. Initializing Pairwise Number k and Session Key

For every billing period, the CC generates randomly the pairwise number for every node in one neighborhood denoted as k, and broadcasts it to all SMs.

We generate session keys between every node with the computational DH key exchange protocol as the initial key in RC4 to generate the keystream between pairwise nodes. Once one node joins a neighborhood size of n, it generates itself one DH public key g^a (mod M) and remains the secret key a, M are DH parameters, and then broadcasts the public key. By this Computational Diffie-Hellman CDH exchange key, any two pairwise nodes can identify their session key formed as g^{ab}.

3.1.2. Modular Addition

The *user$_i$* partitions its own data x_d^i into k partitions denoted as $x_{(ind_i[j],d)}^i$ $(1 \leq j \leq k)$ and sends them to every pairwise node. However, the partitioned data can be easily guessed, especially with brute search, as the consumption value at every timeslot is very small. For this, we impose extra noise (masking keys) which is only known by pairwise nodes themselves on the partitioned data to further secure the individual data.

3.1.3. Noise Addition

Masking keys, as extra noise, are generated by pairwise nodes with PRF at every timestamp. The PRF can be implemented with RC4, the specific process can be referred to the Section 2.2.

3.2. Encryption and Aggregation

3.2.1. Data Encryption

(1). Partition of individual data

Each node randomly partitions its individual data into k partitions and sends them to k pairwise nodes along with the masking keys. The partition form is as follows:

$$x_d^i = \sum_{j \in ind_i[s]} x_{(j,d)}^i (1 \leq s \leq k) \tag{3}$$

(2). Generation of pairwise nodes and masking keys

For any node, it chooses randomly any k nodes in one round as its pairing nodes such that if *user$_i$* selects *user$_j$*, then *user$_j$* also selects *user$_i$*. With the session key between them, the two pairwise nodes generate a common key r from RC4; *user$_i$* adds $r_{(j,d)}^i$ to $x_{(j,d)}^i$, and *user$_j$* adds $r_{(i,d)}^j$ which satisfies:

$$r_{(i,d)}^j = -r_{(j,d)}^i (i \in ind_j[s]; j \in ind_i[s]) \tag{4}$$

For *user$_i$*, the generated noise set at the timestamp d can be denoted as $r_{(ind_i[s],d)}^i$ $(s = 1, 2, \ldots, k)$. Note that in order to facilitate the temporal aggregation, the pairwise key generated by an SM at the Tth timestamp should satisfy the following equation:

$$r_{(i,d)}^j = -r_{(j,d)}^i (i \in ind_j[s]; j \in ind_i[s]) \tag{5}$$

(3). Encryption process

At the timestamp d, *user$_i$* adds the pairwise noise to the partitioned data to generate the encrypted message $E_{(j,d)}^i = x_{(j,d)}^i + r_{(j,d)}^i (j \in ind_i[s])$ to k pairwise nodes seperately as well as receiving the encrypted message they sent. The Figure 2 illustrates an example for spatial and temporal aggregation among pairwise users in multi-region groups.

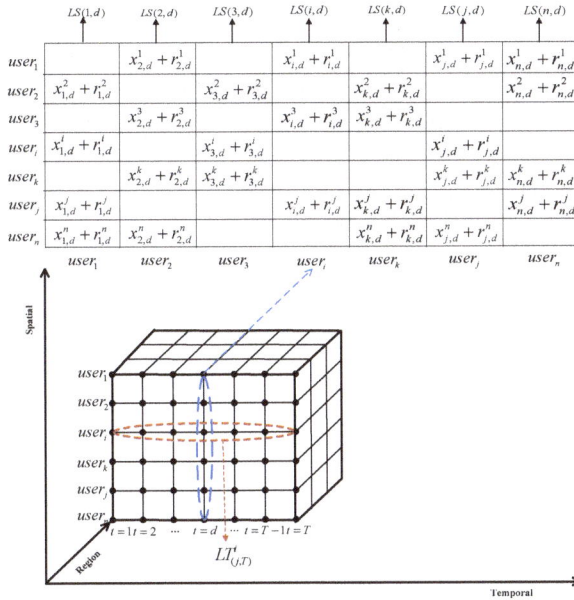

	user$_1$	user$_2$	user$_3$	user$_i$	user$_k$	user$_j$	user$_n$
user$_1$		$x^1_{2,d}+r^1_{2,d}$		$x^1_{i,d}+r^1_{i,d}$		$x^1_{j,d}+r^1_{j,d}$	$x^1_{n,d}+r^1_{n,d}$
user$_2$	$x^2_{1,d}+r^2_{1,d}$		$x^2_{3,d}+r^2_{3,d}$		$x^2_{k,d}+r^2_{k,d}$		$x^2_{n,d}+r^2_{n,d}$
user$_3$		$x^3_{2,d}+r^3_{2,d}$		$x^3_{i,d}+r^3_{i,d}$	$x^3_{k,d}+r^3_{k,d}$		
user$_i$	$x^i_{1,d}+r^i_{1,d}$		$x^i_{3,d}+r^i_{3,d}$			$x^i_{j,d}+r^i_{j,d}$	
user$_k$		$x^k_{2,d}+r^k_{2,d}$	$x^k_{3,d}+r^k_{3,d}$			$x^k_{j,d}+r^k_{j,d}$	$x^k_{n,d}+r^k_{n,d}$
user$_j$	$x^j_{1,d}+r^j_{1,d}$			$x^j_{i,d}+r^j_{i,d}$	$x^j_{k,d}+r^j_{k,d}$		$x^j_{n,d}+r^j_{n,d}$
user$_n$	$x^n_{1,d}+r^n_{1,d}$	$x^n_{2,d}+r^n_{2,d}$			$x^n_{k,d}+r^n_{k,d}$	$x^n_{j,d}+r^n_{j,d}$	

Figure 2. Example multi-region spatial and temporal aggregation.

For any SM node j, it will store the encrypted data sent from one of its pairwise node i in a series of T timestamps in the form of matrix as follows:

$$
\begin{bmatrix}
E^1_{(j,1)} & E^2_{(j,1)} & \cdots & E^i_{(j,1)} & \cdots & E^n_{(j,1)} \\
E^1_{(j,2)} & E^2_{(j,2)} & \cdots & E^i_{(j,2)} & \cdots & E^n_{(j,2)} \\
\vdots & \vdots & & \vdots & & \vdots \\
E^1_{(j,T)} & E^2_{(j,T)} & \cdots & E^i_{(j,T)} & \cdots & E^n_{(j,T)}
\end{bmatrix}
\left(
\begin{array}{ll}
E^i_{(j,1)} = (x^i_{(j,d)}+r^i_{(j,d)}) & j \in ind_i[s] \\
E^i_{(j,d)} = 0 & j \notin ind_i[s]
\end{array}
\right)
\tag{6}
$$

3.2.2. Storage and Aggregation

(1). Spatial Aggregation

Once receiving encrypted data at timeslot d from all pairwise nodes, $user_i$ aggregates them and generates the local spatial data $LS(i,d)$ as follows:

$$
LS(i,d) = \sum_{j \in ind_i[s]}^{n} (x^j_{(i,d)}+r^j_{(i,d)}) \bmod M(1 \leq s \leq k)
\tag{7}
$$

Every user sends the local spatial aggregation formed as $LS(i,d)$ to the GW at every timestamp.

Once receiving the locally spatial aggregation $LS(i,d)$ from the pairwise nodes, the GW adds them up together and the pairwise keys cancel out. The total spatial aggregation is denoted as:

$$
AS_d = \sum_{i=1}^{n} LS(i,d) \bmod M
\tag{8}
$$

(2). Temporal aggregation

Every user node receives the encrypted data from its pairwise nodes and stores it as a matrix of T rows and n columns formed as Equation (6).

In every billing period T, the user node aggregates every column in the Equation (6) into locally temporal aggregation after the pairwise keys cancel out. The locally temporal aggregation form is as follows:

$$
\begin{aligned}
LT^i_{(j,T)} &= \sum_{d=1}^{T} E^i_{(j,d)} \bmod M(j \in ind_i[s]) \\
&= \sum_{d=1}^{T} (x^i_{(j,d)} + r^i_{(j,d)}) \bmod M
\end{aligned}
\tag{9}
$$

Once the CC issues the temporal aggregation request for $user_i$ to the GW, the pairwise nodes of $user_i$ would report its local temporal aggregation $LT^i_{(j,T)}$ to the GW.

The GW aggregates them into the temporal aggregation and transmits it to the CC; the aggregation process is as follows:

$$
AT(i,T) = \sum_{j \in ind_i[s]} LT^i_{(j,T)} \bmod M (1 \le s \le k)
\tag{10}
$$

We assume $j \in ind_i[s]; i \in ind_j[s]$.

Figure 3 shows the communication process between the pairwise nodes and GW at the timestamp d.

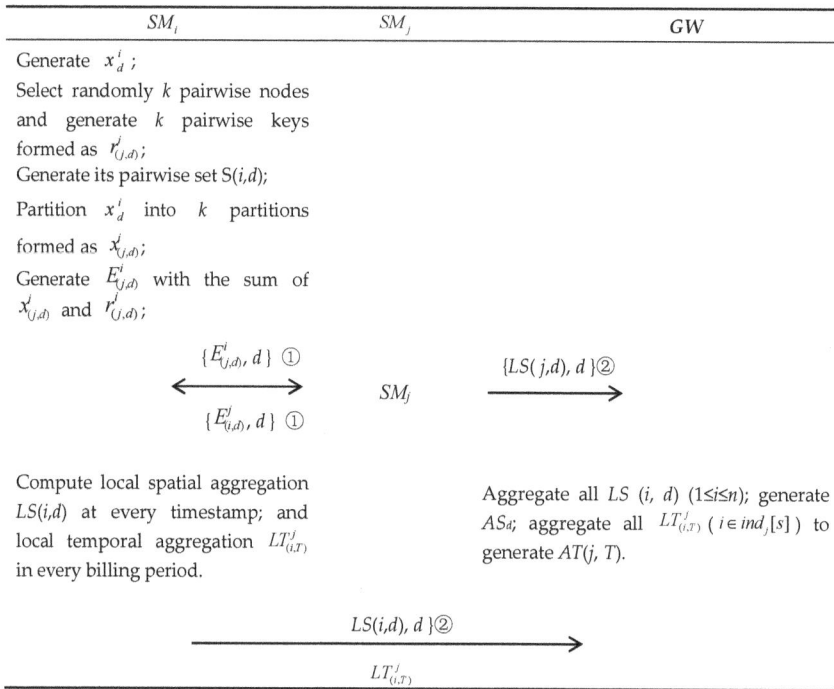

SM_i	SM_j	GW
Generate x^i_d;		
Select randomly k pairwise nodes and generate k pairwise keys formed as $r^i_{(j,d)}$;		
Generate its pairwise set $S(i,d)$;		
Partition x^i_d into k partitions formed as $x^i_{(j,d)}$;		
Generate $E^i_{(j,d)}$ with the sum of $x^i_{(j,d)}$ and $r^i_{(j,d)}$;		
$\{E^i_{(j,d)}, d\}$ ① $\longleftrightarrow$ $\{E^j_{(i,d)}, d\}$ ①	SM_j	$\{LS(j,d), d\}$ ② $\longrightarrow$
Compute local spatial aggregation $LS(i,d)$ at every timestamp; and local temporal aggregation $LT^j_{(i,T)}$ in every billing period.		Aggregate all LS (i, d) $(1 \le i \le n)$; generate AS_d; aggregate all $LT^j_{(i,T)}$ ($i \in ind_j[s]$) to generate $AT(j, T)$.
	$LS(i,d), d$ ② $\longrightarrow$	
	$LT^j_{(i,T)}$	

Figure 3. Communication process data between the pairwise nodes and GW.

3.2.3. Decryption Process

In this way, the aggregation process is actually the decryption process, in which the random keys cancel out and individual consumption in a billing period or the spatial aggregation in a neighborhood is revealed. Hence the combination of simple modular addition with noise addition reduces the costly encryption and decryption operation in public key cryptography.

3.3. Correctness Analysis

Now we prove the correctness of our encryption scheme in terms of spatial and temporal aggregation:

3.3.1. Spatial Aggregation

$$
\begin{aligned}
AS_d &= \sum_{i=1}^{n} LS(i,d) \bmod M \\
&= \sum_{i=1}^{n} \sum_{j \in ind_i[s]} (x^j_{(i,d)} + r^j_{(i,d)}) \bmod M \ (1 \le s \le k) \\
&= (\sum_{i=1}^{n} \sum_{j \in ind_i[s]} x^j_{(i,d)} + \sum_{i=1}^{n} \sum_{j \in ind_i[s]} r^j_{(i,d)}) \bmod M \\
&= (\sum_{j=1}^{n} \sum_{i \in ind_j[s]} x^j_{(i,d)} + \tfrac{1}{2}(\sum_{i=1}^{n} \sum_{j \in ind_i[s]} r^j_{(i,d)} + \sum_{j=1}^{n} \sum_{i \in ind_j[s]} r^j_{(i,d)}) \bmod M \\
&= (\sum_{j=1}^{n} x^j_d + \tfrac{1}{2}(\sum_{i=1}^{n} \sum_{j \in ind_i[s]} r^j_{(i,d)} - \sum_{i=1}^{n} \sum_{j \in ind_i[s]} r^j_{(i,d)}) \bmod M \\
&= \sum_{j=1}^{n} x^j_d \bmod M
\end{aligned} \tag{11}
$$

We prove the correctness of our spatial aggregation by permuting the row and column of data matrix formed as Figure 2. Equation (11) shows that the spatial aggregation in a neighborhood equals to the sum of locally spatial aggregation, i.e., the sum of individual data.

3.3.2. Temporal Aggregation

$$
\begin{aligned}
AT(i,T) &= \sum_{j \in ind_i[s]} LT^i_{(j,d)} \bmod M (1 \le s \le k) \\
&= \sum_{j \in ind_i[s]} \sum_{d=1}^{T} E^i_{(j,d)} \bmod M \\
&= \sum_{d=1}^{T} \sum_{j \in ind_i[s]} (x^i_{(j,d)} + r^i_{(j,d)}) \bmod M \\
&= \sum_{d=1}^{T} x^i_d \bmod M
\end{aligned} \tag{12}
$$

Equation (12) shows that the temporal aggregation for one user node equals to the sum of local temporal aggregation from its pairwise nodes, i.e., the sum of its individual data in a series of timestamps T. It proves further the correctness of our temporal aggregation.

4. Security Notions

4.1. Security Proof

In this section, we mainly elaborate the security properties of our scheme. In particular, based on the security requirement and attack model discussed in Section 2.6, we prove our scheme can ensure the confidentiality of fine-grained meterings for an individual user and the aggregation integrity that the local aggregation, and total aggregation cannot tampered by malicious individual user nodes or the aggregator.

We firstly construct the Individual Metering Indistinguishable (IMI) security game to represent the adversary's actions.

Definition 1. *(IMI security game).*
   Setup: the challenger runs the initialization algorithm and first initializes a group of size n, then generates the system parameter k to the adversary.

Queries: the adversary can not only capture meters' encrypted report but also acquire the encryption and compromise queries until meeting the constraints.

Encrypt: The adversary A chooses $user_i$ and specifies x_d^i to ask for the ciphertext. The challenger returns it the ciphertext $E(x_d^i)$.

Compromise: The adversary A specifies an integer $q \in \{0, 1, \cdots, n\}$. If $q = 0$, the challenger returns the adversary the aggregator' capability, else returns $user_q$'s message.

Challenge. We denote with $\{\overline{C}\}$ the set of the uncompromised users. The adversary selects randomly two meterings $x_d^{i_0}$ and $x_d^{i_1}$ ($i \in \{\overline{C}\}$) at the timestamp d. The challenger flips a random bit $b \in \{0, 1\}$ uniformly and returns the adversary $E(x_d^{i_b})$.

Guess: The adversary outputs a guess $b' \in \{0, 1\}$, and A wins if $b = b'$ with unignorable advantage.

Definition 2. *(IMI security)*

The proposed temporal and spatial aggregation scheme is IMI if no probabilistic polynomial-time adversaries A have more than an ignorable advantage in the IMI security game. The ignorable function for A is as follows:

$$Adv_A = |\Pr[b = b'] - \frac{1}{2}| = 0 \tag{13}$$

Theorem 1. *The proposed encryption scheme is IMI.*

The intuition behind the theorem is any adversary cannot distinguish the encrypted individual metering and the scheme cannot leak any individual user consumption at the dth timestamp.

Proof:

Setup: The challenger initiates the whole system. The challenger generates a group of scale n and pairs number k, and then gives the parameters (n, k) to the adversary.

Queries:

(1). Spatial aggregation

Encrypt: A issues the encryption query with (i, d, x_d^i) to the challenger. The challenger generates the pairwise key $r_{(j,d)}^i (j \in ind_i[s])$ between the pairwise nodes, and imposes it on the randomly partitioned data $x_{(j,d)}^i (j \in ind_i[s])$ to generate the encrypted measure formed as $E(x_d^i) = x_{(j,d)}^i + r_{(j,d)}^i \mod M(j \in ind_i[s])$.

Compromise: A may compromise the aggregator or up to n-1 users in any pairwise set in order to acquire more messages for object users. However, the compromise will encounter restriction when meeting with uncompromised users.

Challenge. For simplifying the proof process and not losing the generalization, we consider the extreme circumstance that $|\overline{c}| = 2$. If the theorem holds for this circumstance, then it holds for $|\overline{c}| > 2$. We assume the user j is the only uncompromised user in $ind_i[s](1 \leq s \leq k)$. The adversary selects the two meterings and gives (i, j, d, x_d^i, x_d^j) to the challenger, the challenger flips a random bit $b \in \{0, 1\}$ uniformly and returns the adversary $E(x_d^i)$ when b = 0, and returns $E(x_d^j)$ when b = 1, and then

$$
\begin{aligned}
E(x_d^i) &= \sum_{l \in ind_i[s]} (x_{(l,d)}^i + r_{(l,d)}^i) \mod M (1 \leq s \leq k) \\
&= (x_{(j,d)}^i + r_{(j,d)}^i + \sum_{c \in \{ind_i[s] - j\}} (x_{(c,d)}^i + r_{(c,d)}^i) \mod M
\end{aligned}
\tag{14}
$$

$$
\begin{aligned}
E(x_d^j) &= \sum_{l \in ind_j[s]} (x_{(l,d)}^j + r_{(l,d)}^j) \mod M (1 \leq s \leq k) \\
&= (x_{(i,d)}^j + r_{(i,d)}^j + \sum_{c \in \{ind_j[s] - i\}} (x_{(c,d)}^j + r_{(c,d)}^j) \mod M
\end{aligned}
\tag{15}
$$

In the Equations (14) and (15), the adversary A cannot solve the two equations at the dth timestamp and gain the exact $x^i_{(j,d)}$ and even if he knows $r^j_{(i,d)} = -r^i_{(j,d)}$, as the two equations have three unknown variables, so it is more impossible for A to acquire x^i_d and x^j_d which ensures the scheme's security.

(2). Temporal aggregation

$$
\begin{aligned}
E\left(\sum_{d=1}^{T} x^i_d\right) &= \sum_{d=1}^{T} \sum_{l\in ind_i[s]} (x^i_{(l,d)} + r^i_{(l,d)}) \bmod M (1 \le s \le k)\\
&= \left(\sum_{d=1}^{T} (x^i_{(j,d)} + r^i_{(j,d)}) + \sum_{d=1}^{T} \sum_{c\in\{ind_i[s]-j\}} (x^i_{(c,d)} + r^i_{(c,d)})\right) \bmod M\\
&= \left((x^i_{(j,T)} + r^i_{(j,T)}) + \sum_{d=1}^{T-1} (x^i_{(j,d)} + r^i_{(j,d)}) + \sum_{d=1}^{T} \sum_{c\in\{ind_i[s]-j\}} (x^i_{(c,d)} + r^i_{(c,d)})\right) \bmod M\\
&= \left(x^i_{(j,T)} + \sum_{d=1}^{T-1} x^i_{(j,d)} + \sum_{d=1}^{T} \sum_{c\in\{ind_i[s]-j\}} (x^i_{(c,d)} + r^i_{(c,d)})\right) \bmod M
\end{aligned}
\tag{16}
$$

$$
E\left(\sum_{d=1}^{T} x^j_d\right) = \left(x^j_{(i,T)} + \sum_{d=1}^{T-1} x^j_{(i,d)} + \sum_{d=1}^{T} \sum_{c\in\{ind_j[s]-i\}} (x^j_{(c,d)} + r^j_{(c,d)})\right) \bmod M
\tag{17}
$$

$\square$

In the Equations (16) and (17), the two equations with four unknown variables make the adversary A impossible to acquire $x^i_{(j,d)}$ or $x^j_{(i,d)}$.

Hence, the encrypted aggregation method can ensure the individual, fine-grained meterings indistinguishable security as long as there is at least one uncompromised user in its pairwise set. Our security properties are based on the randomness of modular addition and stream cipher which is used to blind the individual meterings.

4.2. Security Analysis

We can prove that our proposed solution will withstand the other attacks discussed in Section 2.6 and ensure the integrity of the aggregated data, whether total aggregation or local aggregation.

(1). Eavesdropping resistance

Our proposed scheme supports the openness of communication flow. Whether it is the internal node with access to the communication flow in a community or the external eavesdropper, they can only get the encrypted individual data $(x^i_{(j,d)} + r^i_{(j,d)})$, local aggregation value $(LS(i,d), LT^i_{(j,d)})$ or total aggregation value $(AS(d), AT(i,T))$ sent by GW to CC. However, all of them can not obtain the fine-grained data. We have proved that even if all but one node is compromised, object metering still cannot be leaked. Hence, the proposed encryption method satisfies the security requirement *R1*.

(2). False command from the GW

The GW attempts to obtain object user's meterings by issuing false billing commands in the name of CC, even if he cannot compromise its pairwise nodes. He tries to obtain valuable information from them at any timestamp. However, even so, he can only get the indistinguishable, individual meterings, due to the Equations (14)–(17).

We cannot exclude the possibility that all pairwise keys of $user_i$ at a timestamp are all compromised nodes by the malicious aggregator or external attackers. In this case, the object $user_i$'s privacy is exposed. That is the $user_i$ does not select any one honest node, then the probability is $1 - \left(\frac{k}{n-1}\right)^{n-1-|c|}$. Obviously, the larger the value of $|c|$ is, the smaller the value of k is, and the bigger the probability is. We improve the probability as much as possible and assume $n = 1000$, $k = 30$, and $|c| = 500$ (50% nodes are compromised), and then the probability is 2.47×10^{-7}, so much small probability implies it is almost impossible that one user does not select any one honest node in one timestamp. Even if we fix a bigger pairs period $T = 1$ month, then we would have to cost 38.51 years to acquire individual data.

4.3. Publicly Verifiable Property

The security requirement *R2* given earlier needs to be satisfied with the publicly verifiable property. We provide the public communication flow between nodes in a neighborhood is to ensure the integrity of aggregation data. Any internal node in the community can verify publicly the accuracy of the local aggregation from other nodes and the total aggregation from the GW without compromising the individual fine-grained data. The special public verification process comprises two parties:

4.3.1. Spatial Verification

Based on the public communication flow, any node in the neighborhood can gain the encrypted message formed as $x^i_{(j,d)} + r^i_{(j,d)}$ from the pairwise nodes, and compute its local aggregation formed as $LS(i,d)$ and $LT^i_{(j,d)}$, and thus the total aggregation AS_d and $AT(i,T)$ for the neighborhood can be computed and compared with the reported result from the GW. If the result is questionable, the user can report directly to the CC. With such a supervision, the CC can detect the fraudulent profile of the malicious GW.

4.3.2. Temporal Verification

The public verification method to the spatial aggregation is equally effective to the temporal verification. For any node, one of its pairwise nodes in the neighborhood gain its encrypted message formed as $x^i_{(j,d)} + r^i_{(j,d)}$ in a billing period before computing its local temporal aggregation, and thus its total temporal aggregation is computed and verified by summing up local temporal aggregations from all its pairwise nodes.

Thus, the billing user itself or any user node can verify the accuracy of the billing from the GW without revealing individual fine-grained data. Hence, they can detect if there is a malicious and fraudulent profile of the malicious GW and reports it to the CC in time.

5. Performance Evaluation

We evaluate the performance of the proposed aggregation scheme to assess the overheads. The performance metrics used in our empirical evaluation are defined as follows:

(1) Computation overhead: node's runtime of the proposed scheme in terms of spatial and temporal aggregation.
(2) Communication overhead: the size of a message transmitted between the nodes and GW (number of bits).
(3) Security parameter k: we analyze the impact of the different value of k on the two overheads.

We compare these results against several existing works [23,24] using performance metrics based on Friendly ARM [25] and the library in [17]. By comparison with them, we intend to illustrate our computing and communication advantages in terms of the combination of PRF and modular addition methods adopted, respectively, in the scheme [23] and [24]. Each experiment consists of 50 independent trials and the averaged results of these trials are reported. The computation time required for these tasks is listed in Table 2.

Table 2. Average time for functions.

Notations	Descriptions	Time Cost
C_{add}	Addition	$\approx$0.038 ms
C_{hash}	Hash (100 randoms)	$\approx$0.85 ms
C_{mul}	Multiplication	$\approx$0.013 ms
C_{henc}	Homomorphic encryption	$\approx$2.7 ms
C_{hdec}	Homomorphic decryption	$\approx$0.61 ms
C_{ma}	Hash/Modular addition	$\approx$0.0023 ms
C_{prf}	Pseudorandom function	$\approx$0.074 ms

We fix the number of users at 1 million; the number of C is 10; the number of GW ranges from 1 to 20. Let n denotes a possible number of users in a group, and it ranges from 1 to 5000. We present the impact of a different number of users in the GW and a different value of k (ranging from 1 to 100) on the performance. We also assume, for simplicity, that all SMs can be functioning normally.

5.1. Computation Overhead

(1). Spatial aggregation

Let C_{ma} and C_{prf} denote respectively the cost of Modular addition operation and keys generation operation with PRF, respectively let C_{add} and C_{mul} denote the cost of addition and multiplication operation respectively, and C_{enc} and C_{dec} denote the cost of homomorphic encryption and decryption operation respectively.

In our spatial aggregation scheme, for every node, partitioning individual data into k partitions costs one C_{ma}; generating k pairwise keys costs $k \cdot C_{prf}$; receiving k encrypted messages and adding them up cost $k \cdot C_{add}$, then the computation overhead per node is $C_{ma} + k \cdot C_{prf} + k \cdot C_{add}$ and the total computation overhead per aggregator is $(n-1) \cdot C_{add}$ for aggregating data from n nodes.

In Erkin et al.'s scheme [23], at the dth time step, every hash function cost is C_{hash}, k masking random keys cost is $k \cdot C_{prf}$ and computing total masking keys cost is $2k \cdot C_{add}$, and then encrypting individual data cost is C_{enc}, so the total computation overhead is $C_{hash} + k \cdot C_{prf} + 2k \cdot C_{add} + C_{enc}$.

In Jia et al.'s scheme [24], at the dth time step, the additive secret sharing cost is C_{ss}, k hash functions cost is $k \cdot C_{hash}$, and then k-order polynomial operation is x^k and k matrix multiplication operations cost is $(k^2 + 2k) \cdot C_{mul}$, so the total computation overhead is: $C_{ss} + k \cdot C_{hash} + (k^2 + 2k) \cdot C_{mul}$.

We provide the individual spatial computation overhead comparison in Table 3.

Table 3. Individual spatial computation overhead comparison (msec).

Scheme	Computation Overhead Per Smart Meter
Scheme in [23]	$C_{hash} + k \cdot C_{prf} + 2k \cdot C_{add} + C_{enc}$
Scheme in [24]	$C_{ss} + k \cdot C_{hash} + (k^2 + 2k) \cdot C_{mul}$
Our scheme	$C_{ma} + k \cdot C_{prf} + k \cdot C_{add}$

As described in the related work, the scheme in Reference [23] sets all nodes as communication nodes instead of selecting a limited number of communication nodes as in ours and [22]; however, for convenient comparison, we assume that k communication nodes are selected, which is on the same experiment platform as ours and the scheme in [23]. Even under such relaxation, we can still prove ours is superior in terms of computation and communication cost through the following performance evaluation.

The Figure 4 plots the comparison of spatial computation overhead between our scheme and the schemes in References [23,24] with the value of k increasing. The Figure 4 shows that the three schemes' computation overheads all increase with the value of k increasing, the computation overhead in Reference [23] and ours are lower compared with the scheme in References [24], in which polynomial operation x^k and k matrix multiplication operations generate too much computation overhead with k growing, it has more cost significantly than ours and Erkin et al.'s scheme [23], ours is lower slightly than the scheme in [23], and both of them are close to $O(k) \cdot C_{prf}$.

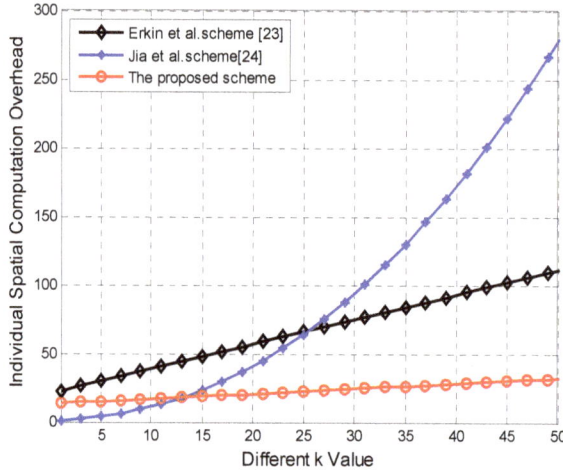

Figure 4. Variety of spatial computation overhead per node with the value k.

(2). Temporal aggregation

In the proposed scheme, each node chooses the same nodes every billing period to satisfy with the Equation (5), so total temporal computation overhead in T serial time slots for every node is $T \cdot (k \cdot C_{prf} + k \cdot C_{add} + C_{ma}) + T \cdot C_{add}$.

In Erkin et al.'s scheme [23], each node sends T fine-grained utility readings in each of the T time steps, so the overhead per node is $T \cdot (C_{hash} + k \cdot C_{prf} + 2k \cdot C_{add} + C_{enc}) + T \cdot C_{mul}$. In fact, the temporal aggregation overhead of the scheme in Reference [23] is higher than it, as with the modification of Paillier encryption, spatial and temporal aggregations are not being synchronized. To compensate the lack, every user must add an additional random key $R_{(i,T+1)} = \dfrac{r^n}{\prod_{d=1}^{T} h_d^{R_{(i,d)}}}$ at T^{th} timestamp, which costs much overhead. However, our scheme has no extra cost and the third party's involvement.

We set the fine-grained reporting interval to be 15 minutes, and billing period $T = 2880$ (roughly one month). Figure 5 plots the comparison of two schemes in terms of temporal computation overhead in every billing period for k ranging from 0 to 50. From Figure 5, we can see the temporal computation overhead per node grows with the increasing of k value in two schemes; however, our proposed scheme increases slightly compared with the scheme in References [23], as the latter costs much overhead on Paillier encryption, while our scheme achieves the same privacy protection effect as the asymmetric encryption with simple and low-cost modular addition.

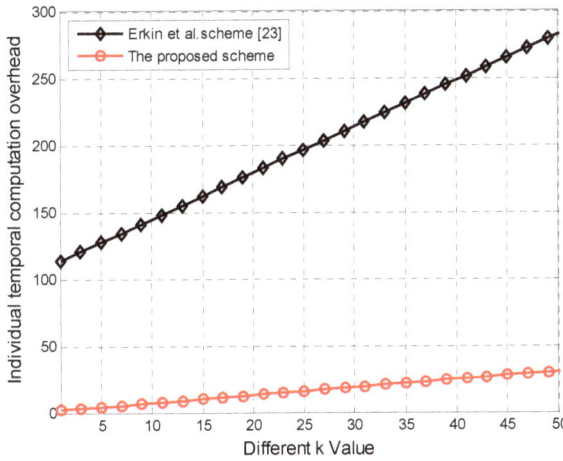

Figure 5. Variety of temporal computation overhead per node with the value k.

5.2. Communication Overhead

We assume the format of a packet is the same as that in TinyOS [26]. The timestamp occupies 128 bits. The sizes of prime numbers p, and q needed in the Paillier encryption are 512 bits each. The size of elements in Z_n^* is 1024 bits. We further assume the plaintext data occupies 32 bits, then random from stream cipher occupies the same byte width with the plaintext data, and Paillier encryption occupies 4096 bits, while the hash function with timestamp occupies 256 bits.

For simplicity, we denote $\{|X|, |R|, |E|, |H|\}$ as the plaintext data size, masking random key (noise) size, Paillier encryption size, and the size of hash function random.

5.2.1. Spatial Communication Overhead Per Node

To generate the spatial aggregation, every node sends the local aggregation to the GW after adding up the encrypted message from all k pairs. The data sent per node can be denoted as $\{LS(i,d)\|t\}$, the size is $|X| + k \cdot |R| + 128$ bits (a partitioned part size is $\frac{|x|}{k}$ bits, k partitions take $|x|$ bits; a noise key takes $|R|$ bits, and then k noise keys take $k \cdot |R|$ bits), so the total packet size is $|x| + k \cdot |R| + 128$ bits.

For the scheme in Reference [23], the spatial aggregation packet per node is in the form as $\{E \| H \| R \| t\}$, its size is $\{|E| + |X| + k \cdot |R| + |H| + 128\}$ bits.

Every user node in Reference [24] generates k results, the data is in the form of $\{(y_1\|y_2\|\cdots\|y_k)\|t\}$, in which y_i involves the computation of data sharing and hash random value, so its size is $K \cdot (K \cdot |H| + |X|/k + 128)$ bits.

We provide the individual spatial communication overhead comparison in Table 4.

Table 4. Individual spatial communication overhead comparison (bits).

Scheme	Computation Overhead Per Smart Meter
Scheme in [23]	$|E| + |X| + k \cdot |R| + |H| + 128$
Scheme in [24]	$K \cdot (K \cdot |H| + |X|/k + 128)$
Our scheme	$|X| + k \cdot |R| + 128$

We plot the individual communication overhead comparison between our scheme and the other two schemes [23,24] during spatial aggregation in the Figure 6. We can see clearly the three schemes' individual overhead all grow with the increasing of k value. The packet width per node in the scheme in Reference [24] grows significantly than the other two schemes, especially when k value is relatively higher, and communication overhead closes to $O(k^2)$, due to the x^k polynomial operation per node before the matrix multiplication operation. Our scheme's growth rate is close to the scheme in Reference [23], which is higher always slightly higher than ours, due to the relatively higher public key encryption width.

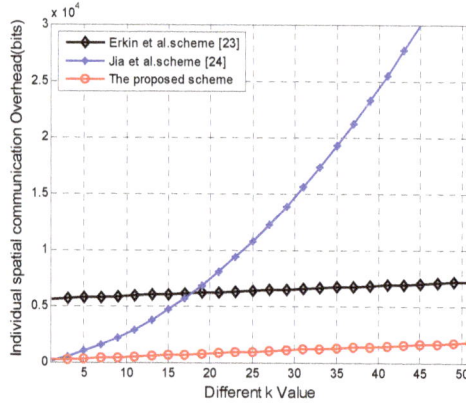

Figure 6. Variety of spatial communication overhead per node with the value k.

5.2.2. Temporal Communication Overhead Per Node

Figure 7 shows the comparison result of ours and the scheme [23] in terms of temporal communication overhead per node when k ranges from 0 to 600, and T ranges from 0 to 6000 mins.

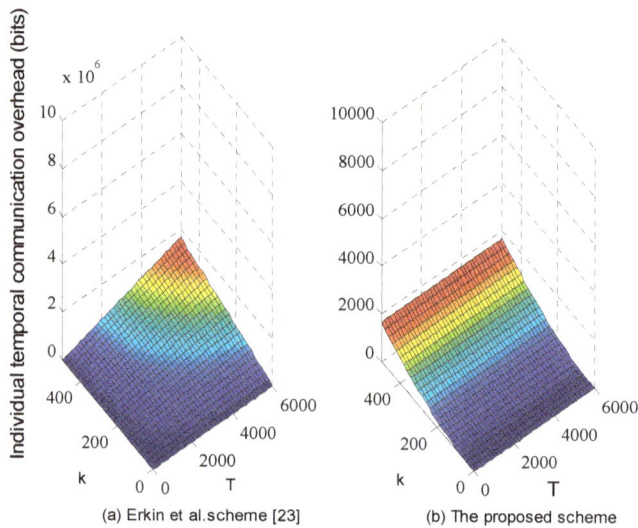

(a) Erkin et al.scheme [23]   (b) The proposed scheme

Figure 7. Variety of temporal communication overhead per node with the values k and T.

In Figure 7, our scheme reduces significantly the packet size sent per node to almost three orders of magnitude than the scheme [23], due to the high overhead of public key encryption. During temporal aggregation, if the process of exchanging random between communication nodes is ignorable, then every node sends its serial encrypted packet formed as $\{E \parallel H \parallel R \parallel t\}(1 \leq t \leq T)$ to the aggregator, so the packet size is $T \cdot (|E| + |X| + k \cdot |R| + |H| + 128)$ bits, while in our scheme, one node's temporal aggregation is computed synchronously before being reported to the aggregator by k communication nodes, and they sends the local temporal aggregation packet size of $|x| + k \cdot |R| + 128$ bits to the aggregator every T timeslot, so aggregating one node's temporal consumption in T serial time slots costs $k \cdot (|x| + k \cdot |R| + 128)$ bits. Hence, when $k \ll T$, ours overhead is always lower significantly lower than the scheme in Reference [23]. Just as the description above, we shorten the number of the communication nodes in Reference [23] into k, and the performance evaluation shows the proposed collection of modular addition and masking keys from PRF saves much computation and communication overhead compared with traditional public key encryption without compromising individual privacy.

6. Conclusions

In the paper, we resolved three issues about privacy-protection aggregation of smart metering customized to the SG. Firstly, the combination of simple modular addition and PRF we designed serves the same effect as the other most related works with lower overhead, namely fending off maliciously internal attacks without compromising individual fine-grained data. Secondly, we proposed innovatively a publicly verifiable platform, by which, every node in a neighborhood can verify local aggregation from every node and total aggregation from the GW and detect the fraudulent profiles from maliciously internal nodes or dishonest user nodes. Thirdly, every node chooses randomly k nodes rather than all nodes as pairwise nodes to communicate, which saves significantly communication and computation overhead, and the independence of the number of users provides scalability and high efficiency under the circumstance of SG big data. From the performance evaluation shows that the proposed scheme is applicable for the security and privacy protection of SG and has practical significance.

Author Contributions: L.Z. and J.Z. designed the hierarchical architecture model, attack models, communication models, and encryption methods together; J.Z. optimized the communication models, and L.Z. wrote the paper.

Funding: This research was funded by National Natural Science Foundation of China (NSFC) (2017–2020, No. 51679058).

Conflicts of Interest: The authors declare no conflicts of interest.

References

1. Wang, W.; Lu, Z. Cyber security in the Smart Grid: Survey and challenges. *Comput. Netw.* **2013**, *57*, 1344–1371. [CrossRef]
2. Ambrosin, M.; Hosseini, H.; Mandal, K. Despicable meter: Anonymous and fine-grained metering data reporting with dishonest meters. In Proceedings of the 2016 IEEE Conference on Communications and Network Security (CNS 2016), Philadelphia, PA, USA, 17–19 October 2016; pp. 163–171.
3. Krebs, B. FBI: Smart Meter Hacks Likely to Spread. 2013. Available online: http://krebsonsecurity.com/2012/04/fbi-smart-meter-hackslikely-to-spread/ (accessed on 07 April 2012).
4. Ohara, K.; Sakai, Y.; Yoshida, F.; Iwamoto, M.; Ohta, K. Privacy-preserving smart metering with verifiability for both billing and energy management. In Proceedings of the 2nd ACM Workshop on ASIA Public-Key Cryptography (ASIAPKC'14), Kyoto, Japan, 3–6 June 2014; pp. 23–32.
5. Lincoln, K.; Philip, K.; Christopher, M. The use of RC4 encryption for smart meters. In Proceedings of the 2014 International Conference on Sustainable Research and Innovation, Nairobi, Kenya, 7–9 May 2014; pp. 58–62.
6. Lu, R.; Liang, X.; Li, X.; Lin, X.; Shen, X. EPPA: An efficient and privacy-preserving aggregation scheme for secure smart grid communications. *IEEE Trans. Parallel Distrib. Syst.* **2012**, *23*, 1621–1631. [CrossRef]
7. Garcia, F.D.; Jacobs, B. Privacy-friendly energy-metering via homomorphic encryption. *IEEE Trans. Parallel Distrib.* **2010**, *6710*, 226–238. [CrossRef]
8. Wang, X.; Mu, Y.; Chen, R. An efficient privacy-preserving aggregation and billing protocol for smart grid. *Secur. Commun. Netw.* **2016**, *9*, 4536–4547. [CrossRef]
9. Chen, L.; Lu, R.; Cao, Z. PDAFT: A privacy-preserving data aggregation scheme with fault tolerance for smart grid communications. *Peer-to-Peer Netw. Appl.* **2015**, *8*, 1122–1132. [CrossRef]

10. He, D.; Kumar, N.; Zeadally, S. Efficient and Privacy-Preserving Data Aggregation Scheme for Smart Grid against Internal Adversaries. *IEEE Trans. Smart Grid* **2017**, *8*, 2411–2419. [CrossRef]
11. Bao, H.; Lu, R. A New Differentially Private Data Aggregation with Fault Tolerance for Smart Grid Communications. *IEEE Internet Things J.* **2015**, *2*, 248–258. [CrossRef]
12. Kursawe, K.; Danezis, G.; Kohlweiss, M. Privacy-friendly aggregation for the smart-grid. In Proceedings of the International Symposium on Privacy Enhancing Technologies Symposium, Cambridge, UK, 27–29 July 2011; pp. 175–191.
13. Shi, Z.; Sun, R.; Lu RChen, L.; Chen, J.; Shen, X.S. Diverse grouping-based aggregation protocol with error detection for smart grid communications. *IEEE Trans. Smart Grid* **2015**, *6*, 2856–2868. [CrossRef]
14. Gupta, S.S.; Maitra, S.; Paul, G.; Sarkar, S. (Non-)Random Sequences from (Non-)Random Permutations—Analysis of RC4 Stream Cipher. *J. Cryptol.* **2014**, *27*, 67–108. [CrossRef]
15. Mahmood, K.; Chaudhry, S.A.; Naqvi, H.; Shon, T.; Ahmad, H.F. A lightweight message authentication scheme for smart grid communications in power sector. *Comput. Electr. Eng.* **2016**, *52*, 114–124. [CrossRef]
16. Li, H.; Lu, R.; Zhou, L.; Yang, B.; Shen, X. An Efficient Merkle-Tree-Based Authentication Scheme for Smart Grid. *IEEE Syst. J.* **2014**, *8*, 655–663. [CrossRef]
17. Chim, T.W.; Yiu, S.M.; Li, V.O.K. PRGA: Privacy-Preserving Recording Gateway-Assisted Authentication of Power Usage Information for Smart Grid. *IEEE Trans. Depend. Secur. Comput.* **2015**, *12*, 85–97. [CrossRef]
18. Fan, C.I.; Huang, S.Y.; Lai, Y.L. Privacy-Enhanced Data Aggregation Scheme Against Internal Attackers in Smart Grid. *IEEE Trans. Ind. Inform.* **2013**, *10*, 666–675. [CrossRef]
19. Castelluccia, C.; Mykletun, E.; Tsudik, G. Efficient Aggregation of encrypted data in Wireless Sensor Networks. In Proceedings of the 2th International Conference on Mobile and Ubiquitous Systems: Networking and Services (MOBIQUITOUS'05), San Diego, CA, USA, 17–21 July 2005; pp. 109–117.
20. Dimitriou, T.; Awad, M.K. Secure and scalable aggregation in the smart grid resilient against malicious entities. *Ad Hoc Netw.* **2016**, *50*, 58–67. [CrossRef]
21. Rahman, M.A.; Manshaei, M.H.; Al-Shaer, E.; Shehab, M. Secure and Private Data Aggregation for Energy Consumption Scheduling in Smart Grids. *IEEE Trans. Depend. Secur. Comput.* **2017**, *14*, 221–234. [CrossRef]
22. Shamir, A. How to Share a Secret. *Commun. ACM* **1979**, *22*, 612–613. [CrossRef]
23. Erkin, Z.; Tsudik, G. Private computation of spatial and temporal power consumption with smart meters. In Proceedings of the 10th International Conference on Applied Cryptography and Network Security (ACNS'12), Singapore, 26–29 June 2012; pp. 561–577.
24. Jia, W.; Zhu, H.; Cao, Z.; Dong, X.; Xiao, C. Human-factor-aware privacy-preserving aggregation in smart grid. *IEEE Syst. J.* **2017**, *18*, 598–607. [CrossRef]
25. FriendlyARM. 2011. Available online: http://www.friendlyarm.net/ (accessed on 17 August 2011).
26. Ahlswede, R.; Csiszar, I. Common randomness in information theory and cryptography I. Secret sharing. *IEEE Trans. Inform. Theory* **1993**, *39*, 1121–1132. [CrossRef]

applied
sciences

MDPI

Article

Smart Appliances for Efficient Integration of Solar Energy: A Dutch Case Study of a Residential Smart Grid Pilot

Cihan Gercek [1,*] and Angèle Reinders [1,2]

[1] Department of Design, Production and Management, Faculty of Engineering Technology, University of Twente, P.O. Box 217, 7500 AE Enschede, The Netherlands; a.h.m.e.reinders@utwente.nl
[2] Energy Technology Group at Mechanical Engineering, Eindhoven University of Technology, P.O. Box 513, 5600 MB Eindhoven, The Netherlands
* Correspondence: c.gercek@utwente.nl; Tel.: +31-534897875

Received: 7 December 2018; Accepted: 28 January 2019; Published: 10 February 2019

Abstract: This paper analyzes the use patterns of a residential smart grid pilot in the Netherlands, called PowerMatching City. The analysis is based on detailed monitoring data measured at 5-min intervals for the year 2012, originating from this pilot which was realized in 2007 in Groningen, Netherlands. In this pilot, smart appliances, heat pumps, micro-combined heat and power (μ-CHP), and solar photovoltaic (PV) systems have been installed to evaluate their efficiency, their ability to reduce peak electricity purchase, and their effects on self-sufficiency and on the local use of solar electricity. As a result of the evaluation, diverse yearly and weekly indicators have been determined, such as electricity purchase and delivery, solar production, flexible generation, and load. Depending on the household configuration, up to 40% of self-sufficiency is achieved on an annual average basis, and 14.4% of the total consumption were flexible. In general, we can conclude that micro-CHP contributed to keep purchase from the grid relatively constant throughout the seasons. Adding to that, smart appliances significantly contributed to load shifting in peak times. It is recommended that similar evaluations will be conducted in other smart grid pilots to statistically enhance insights in the functioning of residential smart grids.

Keywords: smart grids; renewable energy; flexibility; demand shifting; photovoltaic systems; smart appliances

---

1. Introduction

Residential photovoltaic (PV) installations are one of the promising options to locally generate and consume sustainable and cost-effective energy [1]. One of the major technical issues related to the integration of renewable energy systems into local electricity networks is balancing the mismatch between demand and supply of power [2]. Daily and seasonable meteorological conditions significantly affect renewable energy production [3] as well as demand patterns. 100% matching of the residential consumption with renewable energy can be achieved by PV systems in combination with residential storage systems such as batteries [4,5], vehicle to grid technologies [6], or by using community-based storage systems [7]. Although batteries may be required to maintain a high quality of the power fed into local electricity networks, alternative solutions for the realization of flexibility may need to be evaluated because of the high environmental impact of batteries [8]. For instance, one can think about other types of storage systems or optimizing the capacity of batteries.

Rather than self-consumption with flexible loads or temporary storage of the PV infeed to the grid, an efficient and sustainable integration of renewable energy is only possible if the network is flexible and resilient [9]. Electric flexibility can be defined as a power adjustment sustained at a given moment for a given duration from a specific location within the network [10,11]. Thus, a flexibility service is characterized by five attributes: its direction, its electrical composition in power, its temporal characteristics defined by its starting time and duration, and its base for location [10]. To enable all the potential flexibility, the organization and functioning of electricity grids will require more intelligence and complexity, for which reason they are called 'smart grids'. Smart grids balance variations of the energy production in renewable energies with regards to energy demand and regulate the demand side via, for instance, shiftable loads with respect to time and quantity [12].

Firstly, sustainable supply flexibility might be offered by the network itself through storage systems (hydroelectricity, fuel cells, and hydrogen). Hydrogen technologies and fuel cell–powered electric vehicles may provide a balanced energy system [13]; however, such systems are still quite expensive [14]. Therefore, one of the most promising solutions to increase flexibility is the use of combined distributed energy resources (DER), such that they will jointly produce electricity on moments of demand. In this scope, micro-combined heat and power (μ-CHP) units could be complementary to residential PV systems by offering both electricity and heating, especially if the electricity prices are fairly high or natural gas prices are relatively low [15,16]. Therefore, we would like to evaluate the efficient integration of PV systems into a local network that comprises different configurations of DER.

Secondly, residential homes with various smart appliances may contribute to the load flexibility [17] together with home energy management systems and demand response [18,19]. In the literature [20], domestic cleaning practices by use of smart washing machines and smart dishwashers are described as the most favorable residential consumption practices for demand side response [21]. Heating and lighting practices have a medium flexibility potential, according to the same social study. Moreover, in terms of the price responsiveness of electricity users, dishwashers are qualified as significant drivers in time-of-use tariffs [22]. By means of these smart appliances, this study aims to evaluate the flexible load in a smart grid pilot, particularly its temporal characteristics and average electrical composition in power.

In this paper, flexibility in both supply and demand is analyzed by means of detailed monitoring data of PowerMatching City (PMC), which is a residential smart grid pilot which got realized in the year 2007 in the City of Groningen in the Netherlands [23]. This pilot includes 22 households (HH) with PV systems and different configurations of their energy systems with μ-CHP, smart hybrid heat pumps (SHHP), and also smart appliances, as illustrated in Figure 1a and detailed in Table 1 [24]. An energy management software, PowerMatcher, has been used to operate power flows on this pilot [25].

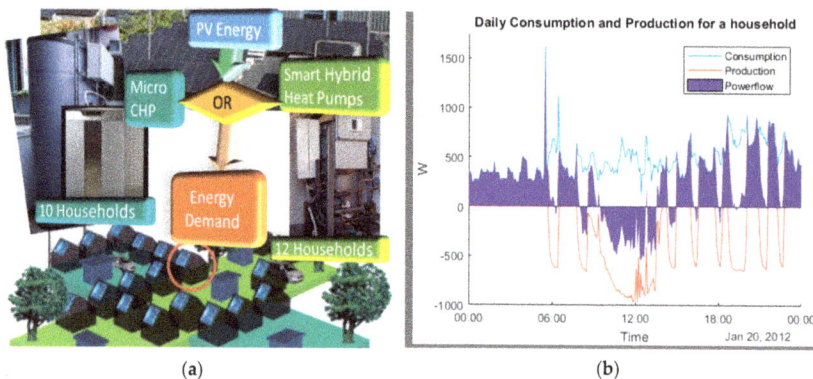

(a)	(b)

Figure 1. PowerMatching City: (**a**) scheme of the system, (**b**) energy consumption (blue line), energy production (red line), and power flow (blue area) for a household with photovoltaic (PV) systems and Micro-combined heat and power (μCHP) jointly producing energy, in the winter of 2012. Negative power flows indicate that power is fed to the grid.

Table 1. Summary of the smart grid pilot settings and photovoltaic (PV) energy production for 2012 in PowerMatching City.

Distributed Energy Sources	Details of PV Systems		PV Production	Smart App.
10 Households with μ-CHP (1kW Electric, 6 KW Thermal)	2 Households: Rooftop PV	500 W_p, 4 m^2 2200 W_p, 18.7 m^2	150 kWh 1835 kWh	No Yes
	8 Households: Virtual PV	1590 W_p, 11 m^2 (on average per household)	880 kWh (on average per household)	4 Households out of 8
12 Households with Hybrid Heat Pump (4.5 kWh Thermal)	2 Households: Rooftop PV	500 W_p, 4 m^2 3840 W_p, 16 m^2	350 kWh 1858 kWh	Yes No
	10 Households: Virtual PV	1590 W_p, 11 m^2 (on average per household)	880 kWh (on average per household)	5 Households out of 10

Our analysis will quantify the electricity consumption of households, purchase of power from the grid, and feed-in of electricity from households to the grid, as well as PV and μ-CHP energy production during 2012, in order to analyze self-sufficiency of residential electricity networks with PV systems in combination with other distributed energy systems. The shiftable load is also quantified to put this analysis in a more complete perspective. In recent EU reports, one of the most dense areas in terms of the investment in smart grids was the Netherlands, and the peak year of the investment was 2012 [26]. The investigated final phase (phase 2) of the pilot duration is January 2012 through January 2015, and the available data for our research was limited to January 2012 until January 2013.

This paper is structured as follows. The pilot configuration, energy tariffs, the data, the data processing methods, data quality, and equations applied to determine energy indicators are presented in Section 2. The results are presented in Section 3 and discussed in Section 4. The paper is completed by conclusions presented in Section 5.

2. Materials and Methods

2.1. PMC Configuration

Main features of PMC households are summarized in Table 1 [24,27]. Ten out of twenty-two households owned a μ-CHP unit with a nominal power of 1 kW of electrical energy. Adding to that, a μ-CHP unit was able to produce 6 kW thermal energy to heat the house, using a hot water buffer of 210 L. Four of the households had rooftop PV installations with an area and nominal power given in Table 1. Through the local smart grid, 18 other households virtually shared a PV system on a farm located 2.3 km from Groningen, the actual location where this smart gird pilot is installed. Each household received a nominal power of 1590 W_p from this farm.

The smart appliances installed in this pilot were smart washing machines, smart dishwashers, and smart hybrid heat pumps (SHHP) with a condensing boiler. The SHHPs contained heat pump units with a thermal power output of 4.5 kW and a condensing boiler with a thermal power output of 20 kW. Additionally, a 210-L hot water buffer was used in these systems. In this pilot, the smart washing machines and smart dishwashers were programmable by time so that users could program it to their needs or comfort expectancies. Half of the households had been equipped with these two smart appliances. The expected outcome of experiments with PMC household smart energy systems matched the time of use of smart appliances with PV power production by a smart algorithm. Figure 1b shows the household consumption and production of a PMC household equipped with PV and μ-CHP in the winter, when irradiance levels are low. It gives an example of how PV and μ-CHP jointly produce electricity to meet the demand, aiming at minimizing the power flow from the grid.

The activation of μ-CHP was organized by PowerMatcher, which not only took into consideration the self-sufficiency of the households but also the neighbors' demand, the dynamic price signal, and the heating requests. Moreover, we were able to see the help of μ-CHP in instantaneously counter-balancing high demands, especially visible in the evening, where usually peak hours occur in the Netherlands (example in Figure 1b is a weekday). The purchase from the grid was brought down to less than 500W by μ-CHP, and the electricity plus including PV production was either delivered to the grid or served for smart appliances, heating, neighbors, etc. Therefore, our aim was to analyze the overall system performance, including DER and smart appliances.

2.2. Household Characteristics

The socio-economic features of households participating in this pilot, such as the education degree and income, was above the average for households in the Netherlands. For instance, the number of residents per household in PowerMatching City (PMC) was 3.1 versus 2.2, which is the average number in the Netherlands [24]. Table 2 gives details about the household sizes. The different shares of single resident households cannot be neglected. Adding to that, the surfaces of the households of PMC were larger than the Dutch average, which may have had an effect on the homes' energy consumption in terms of air conditioning, water heating, and lighting [28]. However, smart appliances and other efficient devices were expected to have a positive effect on the reduction of energy consumption [29].

Table 2. Number of residents per household in PowerMatching City (PMC) vs. the Netherlands [adapted from [24]].

| Household Size | PMC Households | | The Netherlands | |
	Number of Households	Distribution (%)	Number of Households	Distribution (%)
1 person	2	9%	2,761,764	37
2 persons	5	22%	2,455,421	33
3 persons	2	9%	909,274	12
4 persons	7	31%	971,486	13
5 persons	2	9%	414,879	6
Unknown	4	18%	-	-
Total	22	100%	7,512,824	100%

2.3. Energy Tariffs

In the Netherlands, electricity prices for households vary according to the amount of consumption. In 2012, the costs of electricity were [30]:

- 11.5 c€/kWh for 1000–2500 kWh
- 18.7 c€/kWh for 2500–5000 kWh
- 22.3 c€/kWh for 5000–15000 kWh

These numbers include VAT and taxes. Although dynamic prices were applied, we did not have access to real-time prices managed by PowerMatcher [31]. The data meters were also tracking the time-of-use price tariffs, defined as follows. The low-electricity tariff (low rate) was available weekdays from 23:00 to 7:00 and on weekends, beginning on Friday at 23:00 and finishing on Monday morning at 7:00. At other times, the normal electricity tariff (normal rate) applied. The natural gas price was 26 c€/m^3 [32], where 10.6 m^3/m^2 (house surface) was the average Dutch household consumption in 2012.

2.4. Data Processing and Equations

PMC households were equipped with smart meters, which measured the electricity and gas consumption and the production of diverse appliances [33]. Data was stored on a server in the form of cumulative and instantaneous values. Before further processing, the data quality of ~80% of the data was assessed for most of the households for the instantaneous consumption, and values were cross-checked with cumulative values. For the missing 20%, derivatives of cumulative values were employed to fill subsequently missing instantaneous values of more than 15 min to boost the data quality. We excluded only one household with heat pump because of the significantly low data quality, around 50% overall across the year. The power data includes the losses. We processed the 29 variables that were measured for the whole year at a 5-min resolution, using MatLab (2017b). To summarize, these variables consisted of energy delivery and purchase according to two tariffs (shown above in

Section 2.3), PV system power production and μ-CHP power production, the ambient temperature, the status of smart appliances, and several other variables. Table 3 summarizes the equations that have been used to determine energy indicators on the basis of this data set, in which E stands for energy, P for power, t for time, n for number of appliances, m for number of energy sources, n.r. for normal rate, and l.r. for low rate. In this Table 3, to analyze the load flexibility, equations for the status of smart appliances are mentioned in at last three columns, in which S is a logical array of activation time and F is a logical array of flexibility time.

Table 3. Energy indicators and equations used to determine them; E, energy; P, power; t, time; n, number of appliances; m, number of energy sources; n.r., normal rate; l.r., low rate; S, logical array of activation time; and F, logical array of flexibility time.

	Real-Time	Cumulative
Power flow (P_f)	$\sum\limits_{i}^{n} P_{app.i}(t) - \sum\limits_{i}^{n} P_{source\,k}(t)$	$E_P - E_D$
Electricity purchase (E_p)	$P_f(t) > 0$, rate according to t (*see II.C*)	$E_{P\_lr} + E_{P\_n.r.}$
Electricity delivery (E_d)	$P_f(t) < 0$, rate according to t (*see II.C*)	$E_{D\_lr} + E_{D\_n.r.}$
Electricity generated (E_g)	$P_{PV}(t) + P_{\mu\text{-CHP}}(t)$	$E_{PV} + E_{\mu\text{-CHP}}$
Self-consumption (E_s)	$P_{PV}(t) + P_{\mu\text{-CHP}}(t) - \|P_f(t) < 0\|$	$E_G - E_D$
Electricity consumption (E_c)	$\|P_f(t) > 0\| + P_{PV}(t) + P_{\mu\text{-CHP}}(t) - \|P_f(t) < 0\|$	$E_P + E_S$
Smart appliance's activation time (S)	$S_i(t) = 1$, smart appliance running $S_i(t) = 0$, smart appliance is not active	$\sum\limits_{i}^{n} S_i(t)$
Smart appliance's flexibility time (F)	$F_i(t) = 1$, smart appliance is waiting to run $F_i(t) = 0$, smart appliance is not available for flexibility	$\sum\limits_{i}^{n} F_i(t)$
Smart appliance's number of cycles	$S_i(t)$/Average cycle time	$\frac{(\sum\limits_{i}^{n} S_i(t))}{\text{Average cycle time}}$
Smart appliance's electricity consumption	$\sum\limits_{i}^{n} S_i(t) \times$ Average consumtion	Number of cycle × Average consumption per cycle
Heat pump electricity consumption	$P_{hp}(t)$	$\int P_{hp}(t)$

In most of the cases, the consumption of the smart appliances varies significantly with the setting of the program which is used to run them, and their specific consumption curve over time. Hence, the most precise way to determine the power consumption is obviously to measure the consumption of the appliance directly. The power consumption data of the heat pumps helped us to directly obtain this information, processed as in the last row in Table 3.

High-potential flexible loads in this study were dishwashers and washing machines. The power consumption was not measured directly for these two appliances because only smart appliances' activation times (S) were given in the data. To accurately predict the electricity consumption just by activation parameters, two parameters have to be known: the consumption profiles of all use modes that are possible for the specific model of the appliance and its various program (use) cycle times. Moreover, the cycle time of the different programs or combinations of programs of the appliances should not overlap in the time resolution given (here, 5 min). Otherwise, a running program has to be coded as well. Because we did not have information about the specifications of the activated program or the specific consumption profiles and cycle times, we chose to proceed with average values of power consumption within the data set and to consider average cycles. Moreover, the efficiency and energy consumption of the appliances in 2012 are not same as nowadays.

A medium-potential flexible load in this study were heat pumps, as the temperature inside the house had to be kept within the temperature range that users indicated. In fact, indoor temperature is highly dependent on many parameters, such as household orientation, surface, isolation, outside temperature, etc. Adding to that, surveys also indicate a medium potential on behavioral change regarding heating habits [21]. Nevertheless, we still indicate this medium potential of flexible load amount in the results section, including gas consumption in the case that temperatures were too low. This also includes the amount of electricity spent on heating and boiler functions, as the heat pumps provided both. For the performance analysis of the heat pumps and PowerMatcher supervision, please refer to [34].

For the two smart appliances that were analyzed, data from the selected 9 households covered 97.8% of the year on an average. The average cycle of washing machines and dishwashers (2 h and 2 h 15 min, respectively) and the energy consumption per cycle (0.88 kWh and 1.19 kWh respectively) are constants which have been found in the literature, which allowed to calculate the total amount of energy consumed by those two appliances [35,36].

3. Results

From our data analysis it can be seen that the average electricity consumption of the households in PMC in 2012 was 57% higher (5.2 MWh) than the average energy consumption of households in the Netherlands (3.3 MWh) [37]. The higher amount of consumption of PMC can be partially explained by the number of residents per household (40% higher than the Dutch average). PMC households purchased on average 4.3 MWh of electricity. 10 households were able to deliver 0.6MWh of electricity to the grid, while 12 households could not deliver any electricity at all because the configurations of their energy systems did not allow them to produce more than what was consumed at any given moment and because their PV panels were virtual.

On an average, 0.9 MWh of produced energy was self-consumed for households in PMC. Figure 2 illustrates the percentages of energy consumption and production, taking into account different tariffs.

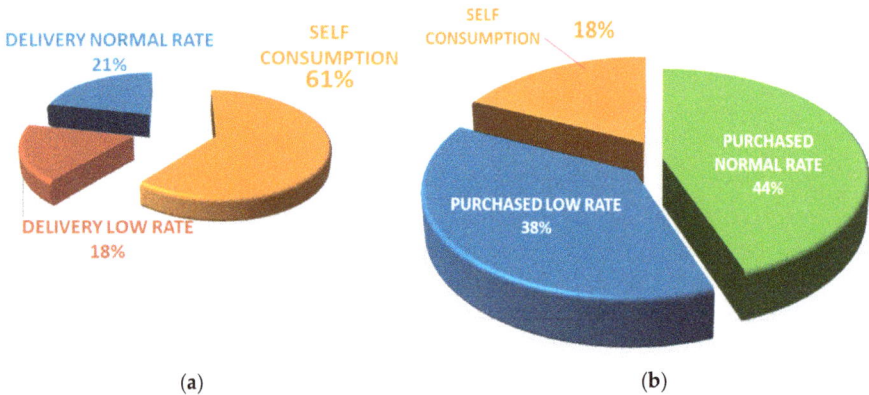

(a) (b)

Figure 2. PowerMatching City electricity purchase, self-consumption, and production in 2012: (a) Electricity generated by PV and µCHP; percentages of self-consumption and electricity delivery depending rates; (b) Self consumption and electricity purchased from the grid, with purchase percentages depending on rates.

In total, all households in PMC produced more than 17 MWh with their PV systems and 9.5 MWh with their µ-CHP units, and 7.2 MWh of electricity was sold, with 42% at a normal tariff. In total, for 21 households, 90 MWh electricity was purchased from the grid with 46% at a low tariff.

3.1. Electricity Consumption and Production Characteristics of PMC Households

Figure 3 and Table 4 provide an energy summary per household for the year 2012, for 5 different energy features mentioned: purchased electricity, electricity production, electricity delivery, self-consumption, and electricity consumption. They vary by up to 400% depending on household size and human behavioral changes from one household to other.

The values are expressed in Table 4, with supplementary values on PV output and µ-CHP. The large range of sample highlights the different groups included in the pilot, which gives more cases and scenarios; however, this might have an effect on average values as the sample is limited to 21 households.

Table 4. Minimum, average, and maximum values for the PMC Households for the year 2012.

kWh/HH		Min	Average	Max
Production	Delivery Low rate	144	350	937
	Delivery Normal Rate	32	247	427
	μ-CHP	292	950	1396
	PV	152	866	1858
	Total Production	350	1277	2832
Consumption	Purchase Low rate	712	1997	3587
	Purchase Normal rate	964	2291	3332
	Self-Consumption	247	935	1626
	Total Consumption	2193	5183	8188

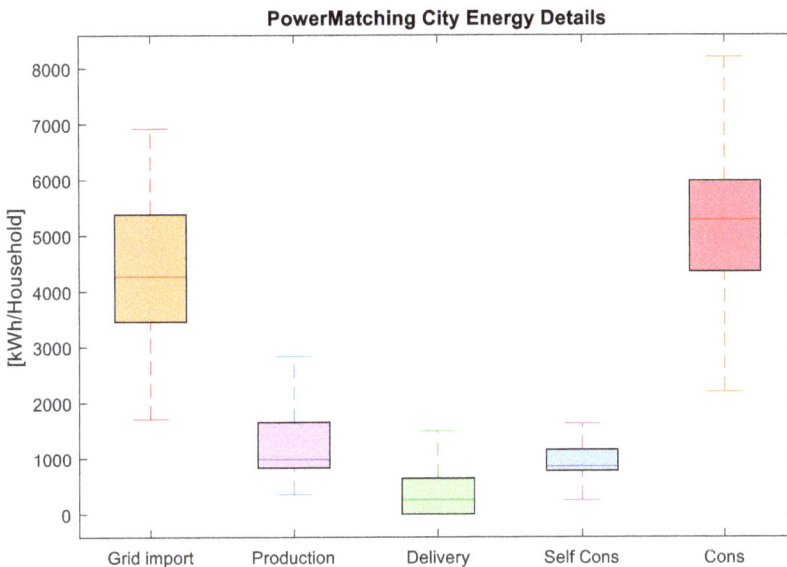

Figure 3. PowerMatching City electricity summary for the year 2012 per household.

3.2. Monthly and Weekly Energy Balance of the Households

As heat pumps were employed in PMC, the electricity consumption obviously increased considerably in winter, and the PV production decreased, as shown in Table 5. In January, the production of μ-CHP was highest and decreased until September (except April). This highest peak seems to be the first trial of μ-CHP for the pilot, as afterwards the values settled between 85 and 105 kWh/household for the winter. Despite the fact that the highest PV production occurred in May, the lowest grid import happened in April, taking into account μ-CHP, otherwise in August for the group of households with only heat pumps.

To focus on the import from the grid, given as averages with higher resolution, Figure 4 shows the weekly variations of electricity purchase, PV, and μ-CHP output in Figure 4a–c, respectively. Figure 4a shows that the average weekly electricity purchase per household (households with HP and μ-CHP) varied slightly throughout the year, except for holiday weeks, as the last week of December and the beginning of the August, where the users were expected to be away. The average value of weekly electricity purchased is 78 kWh per household vs. 96 kWh consumed, which shows the amount of self-consumption. The PV output (Figure 4b) and μ-CHP (Figure 4c) succeeded relatively to keeping the electricity purchase from the grid relatively constant although the seasonal consumption varied a lot in PMC and in most western countries such as the Netherlands.

Table 5. Monthly electricity consumption, production, and grid import on average per household for 12 households with heat pumps and 10 households with µ-CHP.

kWh/HH	Consumption			Production		Grid Import *	
Month	Min.	Mean	Max	PV	µ-CHP	Only PV	PV + µ-CHP
Jan	167	516	824	27	217	489	271
Feb	148	466	633	28	90	438	347
Mar	89	479	561	66	69	413	344
Apr	152	347	645	96	92	251	159
May	99	346	512	137	26	209	183
June	96	326	473	95	20	231	211
July	121	330	708	109	17	221	205
Aug	55	296	509	113	17	183	166
Sep	149	321	637	83	52	238	186
Oct	146	529	570	44	88	485	398
Nov	146	520	579	25	106	495	389
Dec	125	561	640	9	102	552	451

* according to the production subtracted from the mean consumption value.

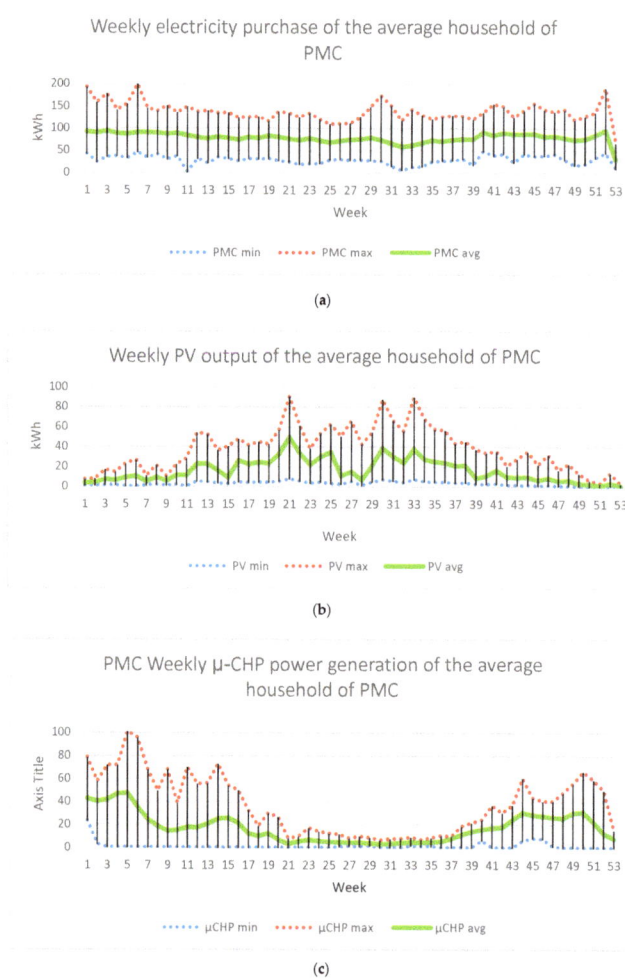

(a)

(b)

(c)

Figure 4. PowerMatching City (PMC) weekly values of (**a**) electricity purchase, (**b**) PV output, and (**c**) µ-CHP (micro-combined heat and power) power generation.

3.3. Flexibility with Smart Appliances

On average per household, 408 h of dishwasher activity and 297 h of washing machine activity were recorded during the year of 2012. By using the equations mentioned in Table 2, the average consumption of the washing machine per household in 2012 is 130.83 kWh. This value is 10% lower than the average Dutch laundry electricity consumption and 25% lower than in the EU-15 [35], which clearly shows the efficient use of smart washing machines comparing to classical machines, despite the greater number of household residents. The same methodology is applied for the smart dishwasher. The total consumption of the dishwasher for 2012 was 215.7 kWh per household. This value is 8% lower than the average value for the EU-15 [36]. In total, these smart appliances brought 346.5 kWh of load flexibility to the grid per household, which corresponds to 10.5% of the average Dutch household electricity consumption in 2012, and to 6.8% of the average household electricity consumption in PMC. Table 6 summarizes the average consumption of smart appliances based on the equations presented in Table 2 and compares those consumptions to average values of different countries.

Table 6. Smart appliance electrical consumption.

| Smart Appliance Type | PowerMatching City (2012) | | Comparing to | |
	Data Fraction (%)	Annual Energy Consumption	Dutch Average [6] (%)	EU-15 Average [7] (%)
Dishwasher	97.8%	215.7 kWh	N.A.	−8%
Washing Machine	95.2%	130.8 kWh	−10%	−25%

4 out of 9 washing machine data were recorded without any activation during a considerable period, making their consumption abnormally close to zero, they were excluded from the analysis.

On average per household, 408 h of dishwasher activity and 297 h of washing machine activity were recorded during the year of 2012. By using the equations mentioned in Table 2, the average consumption of the washing machine per household in 2012 is 130.83 kWh. This value is 10% lower than the average Dutch laundry electricity consumption and 25% lower than in the EU-15 [35], which clearly shows the efficient use of smart washing machines comparing to classical machines, despite the greater number of household residents. The same methodology is applied for the smart dishwasher. The total consumption of the dishwasher for 2012 was 215.7 kWh per household. This value is 8% lower than the average value for the EU-15 [36]. In total, these smart appliances brought 346.5 kWh of load flexibility to the grid per household, which corresponds to 10.5% of the average Dutch household electricity consumption in 2012, and to 6.8% of the average household electricity consumption in PMC. Table 6 summarizes the average consumption of smart appliances based on the equations presented in Table 2 and compares those consumptions to average values of different countries.

Those percentages should be considered with the semi-automatic flexibility that it could contribute by users entering a latest run time. By using the same equations presented in Table 2, we considered this time how much the two smart appliances were shifted compared to their use time. The flexibility period offered by smart washing machines (WM) was similar to the usage time in median value and 60% for the smart dishwasher (DW) in comparison to the usage time (Figure 5).

Defining peak hours as 17:00–23:00 on weekdays, we found that on average, 60% of the DW peak hours were shifted with the latest runtime option and, similarly, 20% for WM, comparing again the actual runtime in the same year.

As providing full-automatic flexibility, the heat pumps with electric water boiler option consumed 400 kWh on average per household. Heating was mainly done by gas consumption, which is detailed in the discussion.

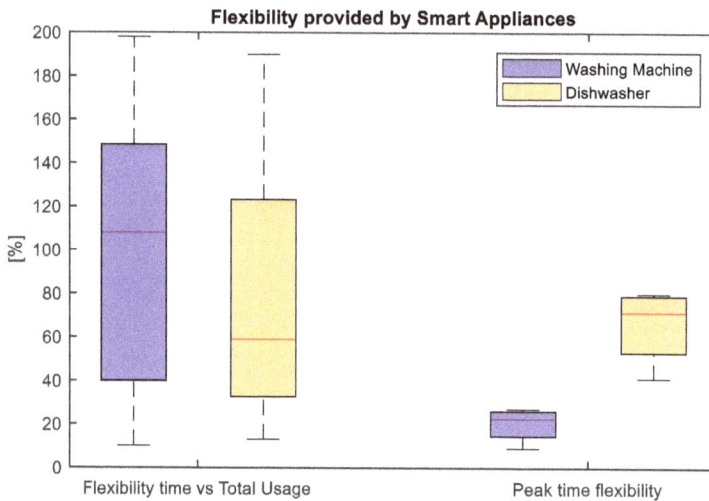

Figure 5. PowerMatching City smart washing machine and dishwasher. The latest run time program was compared to the actual use in 2012, shown as total flexibility time and flexibility offered during peak time.

4. Discussion

As the number of residents, their socio-economic status, educational level, and the surface of the households are above the Dutch averages, and the sample is limited to 21 households and one single year, results should be considered with caution with respect to the average values. However, these results indicate to which extend PV infeed can be applied and how much self-sufficiency or energy savings may be obtained with the combination of technologies such as µ-CHP and heat pumps, knowing that using only PV will not be appropriate for the grid in northern countries such as the Netherlands. Yearly and weekly results highlight the benefits of DER combination in the seasonal variabilities of renewable energies. In this section, we present a simplistic energy bill analysis and discuss the flexibility before stating our conclusions.

4.1. PV and µ-CHP: Electricity Production

Figure 4b indicates the weekly PV output in mean value with an average of 37 kWh, which is the same for µ-CHP power generation, shown in Figure 4c. The winter atmospheric conditions in the Netherlands have a consequent impact on the PV output, and µ-CHP power generation shows a good way of balancing the seasonal renewable energy output. This energy comes at an additional cost of gas consumption, which can be provided from sustainable sources such as biogas.

Gas consumption was 1700 m^3, which is higher than the Dutch average (1400 m^3) [38], but considering the surfaces varying between 150 and 199 m^2 in the PMC, this values gets close to the Dutch average: 10.6 m^3/m^2 household surface. As the PV and µ-CHP provided a part of the consumption, the households could decrease their bill considerably, especially when they were potentially purchasing one electricity rate category lower. µ-CHP gas consumption costs 390€/year on average, similar to other heating means, and helps to maintain the seasonal electricity balance. Moreover, in time of use or dynamic tariff scenarios, their role will be much more important, as the peak hour purchase will vary.

4.2. Self Sufficiency

The households equipped with PV and µ-CHP had a self-sufficiency of 40.7% over the year, contrary to those furnished with PV and SHHP (20.3%). As presented in Figure 4, even in the worst weeks, µ-CHP had a self-sufficiency of over 20%, except for one week when it dropped to 12%. SHHP, which drops rapidly to under 5% for many winter weeks, even saw 1%, as it was only related to PV generation.

4.3. Energy Bill

To simplify and not take into account different prices and tariffs that different electricity providers offer, we will use the most common tariff in the Netherlands, net metering. The installation costs will be excluded from our analysis, as such information is difficult to access in hindsight. Only data on the difference between the total consumption and total purchase, and what is saved from the electricity bill has been considered.

For obvious reasons, the installation capacities of PV Wp played a major role in the savings. Up to 640 €/year were saved with μ-CHP and PV installations, with mean savings of 510 €/year in this group. The group without μ-CHP but with PV infeed saved 336 €/year on average, and the minimum savings were 40 €/year, because of a very small PV installation. Although savings are quite considerable, the real installation costs and maintenance costs have to be taken into consideration in order to conclude the real economic benefits of the smart grid for the prosumers.

4.4. Flexibility and Smart Washing Machine and Dishwasher

We observed that, despite the high number of residents per house and an energy consumption which was higher than the national average, the smart appliances in the PMC pilot were consuming less energy than the traditional ones. However, the arguments are not strong enough to draw the conclusion that smart appliances reduce the energy consumption, as the number of households was too low to be statistically significant (11). Additionally, there might be a bias in user behavior and the activation frequency of the appliances due to the Hawthorne effect (see [39]). More multidisciplinary studies as mentioned in [12] on the subject are needed, especially regarding user behaviors [40–42].

Current machines have become much more efficient, and accordingly, the residential electricity consumption has been decreasing since 2012. To respect the conditions at the time of data collection and to analyze the ratio between the overall electricity consumption and the smart appliances' role, we have chosen literature from the same period, which we estimated being more significant in percentages. The amount of electricity and the existing ratio should be approved with current machines and ongoing smart grid initiatives.

5. Conclusions

To sum up, we compared the different configurations of a smart grid pilot in the Netherlands in order to identify the demand–supply balance of the configurations. Energy consumption and energy production is classified in three groups: low rate, normal rate, and energy self-consumption. The weekly purchase, PV output, and μ-CHP power generation is shown to highlight the seasonal complexity. The impact on the energy bill is discussed in the previous section, as well as the flexibility provided by smart appliances and heat pumps, which together corresponds to 14.4% of the electricity consumption for an average household.

In conclusion, μ-CHP might be a good solution for northern countries such as the Netherlands to provide heat and electricity when PV infeed is weak. The installation costs and the complexity to integrate this kind of equipment in existing buildings, as well as the insulation class of the household might be the barriers in those configurations. Regarding flexibility contributions, we support the findings of the social scientists that cleaning practices are potentially highly flexible for residential consumptions, which we demonstrated in this work to be as flexible as their usage time across the whole year.

Author Contributions: Writing—original draft preparation, formal analysis, investigation and algorithms, C.G.; conceptualization and validation, C.G. and A.R.; resources, review, editing, A.R. and C.G.; supervision, project administration, data and funding acquisition, A.R.

Appl. Sci. **2019**, *9*, 581

Funding: This research has received funding from the European Union's Horizon 2020 research and innovation programme under the ERA-Net Smart Grids plus, grant number 646039, from the Netherlands Organisation for Scientific Research (NWO).

Acknowledgments: Our project has received funding in the framework of the joint programming initiative ERA-Net Smart Grids Plus, with support from the European Union's Horizon 2020 research and innovation programme. Furthermore, we would like to acknowledge all participants in the smart grid pilot (PowerMatching City) involved in this study for their willingness to share their data, experiences, and knowledge with the researchers.

Conflicts of Interest: The authors declare no conflict of interest.

Disclaimer: The content and views expressed in this material are those of the authors and do not necessarily reflect the views or opinion of the ERA-Net SG+ initiative. Any reference given does not necessarily imply the endorsement by ERA-Net SG+.

References

1. van Geenhuizen, M.; Schoonman, J.; Reinders, A. Diffusion of Solar Energy Use in the Built Environment Supported by New Design. *J. Civ. Eng. Archit.* **2014**, *8*, 253–260. [CrossRef]
2. Campbell, M. The Economics of PV Systems. In *Photovoltaic Solar Energy*; Reinders, A., Verlinden, P., van Sark, W., Freundlich, A., Eds.; John Wiley & Sons Ltd.: Chichester, UK, 2017; pp. 621–633, ISBN 978-1-118-92749-6.
3. Reinders, A.; Verlinden, P.; van Sark, W.; Freundlich, A. *Photovoltaic Solar Energy: From Fundamentals to Applications*; John Wiley & Sons Ltd.: Chichester, West Sussex, UK; Hoboken, NJ, USA, 2017; ISBN 978-1-118-92746-5.
4. Quintero Pulido, D.; Hoogsteen, G.; ten Kortenaar, M.; Hurink, J.; Hebner, R.; Smit, G. Characterization of Storage Sizing for an Off-Grid House in the US and the Netherlands. *Energies* **2018**, *11*, 265. [CrossRef]
5. Schram, W.L.; Lampropoulos, I.; van Sark, W.G.J.H.M. Photovoltaic systems coupled with batteries that are optimally sized for household self-consumption: Assessment of peak shaving potential. *Appl. Energy* **2018**, *223*, 69–81. [CrossRef]
6. Hoogvliet, T.W.; Litjens, G.B.M.A.; van Sark, W.G.J.H.M. Provision of regulating- and reserve power by electric vehicle owners in the Dutch market. *Appl. Energy* **2017**, *190*, 1008–1019. [CrossRef]
7. AlSkaif, T.; Schram, W.; Litjens, G.; van Sark, W. Smart charging of community storage units using Markov chains. In Proceedings of the Innovative Smart Grid Technologies Conference Europe, Torino, Italy, 26–29 September 2007; IEEE: Torino, Italy, 2017; pp. 1–6.
8. Peters, J.F.; Baumann, M.; Zimmermann, B.; Braun, J.; Weil, M. The environmental impact of Li-Ion batteries and the role of key parameters—A review. *Renew. Sustain. Energy Rev.* **2017**, *67*, 491–506. [CrossRef]
9. Kabalci, E. Power System Flexibility and Resiliency. In *Power Systems Resilience*; Mahdavi Tabatabaei, N., Najafi Ravadanegh, S., Bizon, N., Eds.; Springer International Publishing: Cham, Switzerland, 2019; pp. 81–100, ISBN 978-3-319-94441-8.
10. Eid, C.; Codani, P.; Perez, Y.; Reneses, J.; Hakvoort, R. Managing electric flexibility from Distributed Energy Resources: A review of incentives for market design. *Renew. Sustain. Energy Rev.* **2016**, *64*, 237–247. [CrossRef]
11. Lannoye, E.; Flynn, D.; O'Malley, M. Evaluation of Power System Flexibility. *IEEE Trans. Power Syst.* **2012**, *27*, 922–931. [CrossRef]
12. Reinders, A.; Übermasser, S.; van Sark, W.; Gercek, C.; Schram, W.; Obinna, U.; Lehfuss, F.; van Mierlo, B.; Robledo, C.; van Wijk, A. An Exploration of the Three-Layer Model Including Stakeholders, Markets and Technologies for Assessments of Residential Smart Grids. *Appl. Sci.* **2018**, *8*, 2363. [CrossRef]
13. Oldenbroek, V.; Verhoef, L.A.; van Wijk, A.J.M. Fuel cell electric vehicle as a power plant: Fully renewable integrated transport and energy system design and analysis for smart city areas. *Int. J. Hydrogen Energy* **2017**, *42*, 8166–8196. [CrossRef]
14. Robledo, C.B.; Oldenbroek, V.; Abbruzzese, F.; van Wijk, A.J.M. Integrating a hydrogen fuel cell electric vehicle with vehicle-to-grid technology, photovoltaic power and a residential building. *Appl. Energy* **2018**, *215*, 615–629. [CrossRef]

15. Salmerón Lissén, J.; Romero Rodríguez, L.; Durán Parejo, F.; Sánchez de la Flor, F. An Economic, Energy, and Environmental Analysis of PV/Micro-CHP Hybrid Systems: A Case Study of a Tertiary Building. *Sustainability* **2018**, *10*, 4082. [CrossRef]

16. Beccali, M.; Ciulla, G.; Di Pietra, B.; Galatioto, A.; Leone, G.; Piacentino, A. Assessing the feasibility of cogeneration retrofit and district heating/cooling networks in small Italian islands. *Energy* **2017**, *141*, 2572–2586. [CrossRef]

17. Staats, M.R.; de Boer-Meulman, P.D.M.; van Sark, W.G.J.H.M. Experimental determination of demand side management potential of wet appliances in the Netherlands. *Sustain. Energy Grids Netw.* **2017**, *9*, 80–94. [CrossRef]

18. Zhai, S.; Wang, Z.; Yan, X.; He, G. Appliance Flexibility Analysis Considering User Behavior in Home Energy Management System Using Smart Plugs. *IEEE Trans. Ind. Electron.* **2019**, *66*, 1391–1401. [CrossRef]

19. Godina, R.; Rodrigues, E.; Pouresmaeil, E.; Matias, J.; Catalão, J. Model Predictive Control Home Energy Management and Optimization Strategy with Demand Response. *Appl. Sci.* **2018**, *8*, 408. [CrossRef]

20. Weck, M.H.J.; van Hooff, J.; van Sark, W.G.J.H.M. Review of barriers to the introduction of residential demand response: A case study in the Netherlands: Barriers to residential demand response in smart grids. *Int. J. Energy Res.* **2017**, *41*, 790–816. [CrossRef]

21. Smale, R.; van Vliet, B.; Spaargaren, G. When social practices meet smart grids: Flexibility, grid management, and domestic consumption in The Netherlands. *Energy Res. Soc. Sci.* **2017**, *34*, 132–140. [CrossRef]

22. Guo, P.; Lam, J.C.K.; Li, V.O.K. Drivers of domestic electricity users' price responsiveness: A novel machine learning approach. *Appl. Energy* **2019**, *235*, 900–913. [CrossRef]

23. PowerMatching City A Demonstration Project of a Future Energy Infrastructure. Available online: http://powermatchingcity.nl/ (accessed on 12 April 2018).

24. Geelen, D.V. Empowering End-Users in the Energy Transition: An Exploration of Products and Services to Support Changes in Household Energy Management. Ph.D. Thesis, TU Delft, Delft, The Netherlands, 2014.

25. Kok, K. The powermatcher: Smart coordination for the smart electricity grid. Ph.D. Thesis, Vrije University, Amsterdam, The Netherlands, 2013.

26. Gangale, F.; Vasiljevska, J.; Covrig, C.F.; Mengolini, A.; Fulli, G. *Smart Grid Projects Outlook 2017: Facts*; Publications Office of the European Union: Luxembourg, 2017; ISBN 978-92-79-68899-7.

27. Geelen, D.; Scheepens, A.; Kobus, C.; Obinna, U.; Mugge, R.; Schoormans, J.; Reinders, A. Smart energy households' pilot projects in The Netherlands with a design-driven approach. In Proceedings of the IEEE PES ISGT Europe 2013, Lyngby, Denmark, 6–9 October 2013; IEEE: New Jersey, NJ, USA, 2013; pp. 1–5.

28. Glasgo, B.; Hendrickson, C.; Azevedo, I.M.L. Using advanced metering infrastructure to characterize residential energy use. *Electr. J.* **2017**, *30*, 64–70. [CrossRef]

29. Yuce, B.; Rezgui, Y.; Mourshed, M. ANN–GA smart appliance scheduling for optimised energy management in the domestic sector. *Energy Build.* **2016**, *111*, 311–325. [CrossRef]

30. Centraal Bureau voor de Statistiek. *Elektriciteit in Nederland*; Centraal Bureau voor de Statistiek: The Hague, The Netherlands, 2015.

31. Kok, K.; Roossien, B.; MacDougall, P.; Van Pruissen, O.; Venekamp, G.; Kamphuis, R.; Laarakkers, J.; Warmer, C. Dynamic pricing by scalable energy management systems—Field experiences and simulation results using PowerMatcher. In Proceedings of the 2012 IEEE Power and Energy Society General Meeting, San Diego, CA, USA, 22–26 July 2012; IEEE: San Diego, CA, USA, 2012; pp. 1–8.

32. Netherlands Enviromental Assessment Agency. *Dutch National Energy Outlook (NEV) 2015*; Netherlands Enviromental Assessment Agency: The Hague, The Netherlands, 2015.

33. Geelen, D.; Brezet, H.; Keyson, D.; Boess, S. Gaming for energy conservation in households. In Proceedings of the Knowledge Collaboration & Learning for Sustainable Innovation, Delft, The Netherlands, 25–29 October 2010; TU Delft: Delft, The Netherlands, 2010; pp. 1–18.

34. Bliek, F.; van den Noort, A.; Roossien, B.; Kamphuis, R.; de Wit, J.; van der Velde, J.; Eijgelaar, M. PowerMatching City, a living lab smart grid demonstration. In Proceedings of the IEEE PES, Gothenberg, Sweden, 11–13 October 2010; IEEE: Gothenberg, Sweden, 2010; pp. 1–8.

35. Pakula, C.; Stamminger, R. Electricity and water consumption for laundry washing by washing machine worldwide. *Energy Effic.* **2010**, *3*, 365–382. [CrossRef]

36. Stamminger, R.; Broil, G.; Pakula, C.; Jungbecker, H.; Braun, M.; Rüdenauer, I.; Wendker, C. *Synergy Potential of Smart Appliances*; Report of the Smart-A Project; University of Bonn: Bonn, Germany, 2008.

37. Energieonderzoek Centrum Nederland. *Energie Trends 2012*; ECN: Petten, The Netherlands, 2012.
38. Centraal Bureau voor de Statistiek. Available online: https://www.cbs.nl/en-gb/news/2014/16/energy-consumption-marginally-down-in-2013 (accessed on 7 December 2018).
39. Schwartz, D.; Fischhoff, B.; Krishnamurti, T.; Sowell, F. The Hawthorne effect and energy awareness. *Proc. Natl. Acad. Sci. USA* **2013**, *110*, 15242–15246. [CrossRef] [PubMed]
40. Reinders, A.; de Respinis, M.; van Loon, J.; Stekelenburg, A.; Bliek, F.; Schram, W.; van Sark, W.; Esteri, T.; Uebermasser, S.; Lehfuss, F.; et al. Co-evolution of smart energy products and services: A novel approach towards smart grids. In Proceedings of the 2016 Asian Conference on Energy, Power and Transportation Electrification (ACEPT), Singapore, 25–27 October 2016; IEEE: New Jersey, NJ, USA, 2016; pp. 1–6.
41. Geelen, D.; Vos-Vlamings, M.; Filippidou, F.; van den Noort, A.; van Grootel, M.; Moll, H.; Reinders, A.; Keyson, D. An end-user perspective on smart home energy systems in the PowerMatching City demonstration project. In Proceedings of the IEEE PES ISGT Europe 2013, Lyngby, Denmark, 6–9 October 2013; IEEE: Lyngby, Denmark, 2013; pp. 1–5.
42. Obinna, U.; Joore, P.; Wauben, L.; Reinders, A. Preferred attributes of home energy management products for smart grids—Results of a design study and related user survey. *J. Des. Res.* **2018**, *16*, 99–130. [CrossRef]

applied
sciences

MDPI

Article

Provision of Ancillary Services from an Aggregated Portfolio of Residential Heat Pumps on the Dutch Frequency Containment Reserve Market

Joeri Posma *, Ioannis Lampropoulos, Wouter Schram and Wilfried van Sark

Copernicus Institute of Sustainable Development, Utrecht University, Princetonlaan 8a,
3584 CB Utrecht, The Netherlands; i.lampropoulos@uu.nl (I.L.); w.l.schram@uu.nl (W.S.);
W.G.J.H.M.vanSark@uu.nl (W.v.S.)
* Correspondence: Joeri149@gmail.com; Tel.: +31631956887

Received: 28 December 2018; Accepted: 5 February 2019; Published: 11 February 2019

Featured Application: Operating an aggregation of heat pumps for market-based provision of Frequency Containment Reserves.

Abstract: This study investigates the technical and financial potential of an aggregation of residential heat pumps to deliver demand response (DR) services to the Dutch Frequency Containment Reserve (FCR) market. To determine this potential, a quantitative model was developed to simulate a heat pump switching process. The model utilizes historical frequency and heat pump data as input to determine the optimal weekly bid size considering the regulations and fine regime of the FCR market. These regulations are set by the Dutch Transmission System Operator (TSO). Two strategies were defined that can be employed by an aggregator to select the optimal bid size; the 'always available' scenario and the 'always reliable' scenario. By using the availability and reliability as constraints in the model, the effects of TSO regulations on the potential for FCR are accurately assessed. Results show a significant difference in bid size and revenue of the strategies. In the 'always available' scenario, the average resultant bid size is 1.7 MW, resulting in €0.22 revenue per heat pump (0.5kW$_p$) per week. In the 'always reliable' scenario, the average resultant bid size is 7.9 MW, resulting in €1.00 revenue per heat pump per week on average in the period 03-10-2016–24-04-2017. This is based on a simulation of 20,000 heat pumps with a total capacity of 1 MWp. Results show a large difference in potential between the two strategies. Since the strategies are based on TSO-regulations and strategic choices by the aggregator, both seem to have a strong influence on the financial potential of FCR provision. In practice, this study informs organizations that provide FCR with knowledge about different bidding strategies and their market impact.

Keywords: demand response; aggregator; heat pumps; FCR; frequency containment reserve; ancillary services; smart grids

---

1. Introduction

With the transition towards a low-carbon energy supply system underway, the share of electricity generated by renewable energy resources (RER) is likely to increase. In Europe, wind and solar energy have the highest potential in terms of renewable electricity generation [1]. Wind and solar energy resources are intermittent, as their availability depends on weather patterns [2]. To maintain the system frequency within acceptable limits, electricity supply and demand needs to be balanced. Traditionally, supply could be adjusted to match demand by dispatching fossil-fueled generators. Given the transition towards renewable energy generation, the traditional electricity grid needs to be shifted towards the so-called *smart grid* notion, in which new technologies can be implemented without jeopardizing grid reliability and efficiency, and which make the grid less environmentally friendly [3]. A smart grid allows grid-elements which are only passively used in the current electricity system to become actively involved in the provision of system services such as balancing activities [4]. An important aspect of smart grids is Demand Response (DR). Also known as a category of the general term demand side management [5,6], Demand response is defined by [7] as 'The process through which final consumers (households or businesses) provide flexibility to the electricity system by voluntarily changing their usual electricity consumption in reaction to price signals or to specific requests, while at the same time benefiting from doing so'.

In the deregulated energy market, a consumer that takes an active role in energy generation and/or provision of flexibility services is referred to as a prosumer [8,9]. For small consumers to become prosumers by using their flexible load for financial or balancing purposes, a new role is needed in the energy value chain: the role of aggregators. An aggregator is defined by [10] as 'an intermediary between small consumers and other players (e.g., the retailers, or distribution companies) in the system' (pp.138). Aggregators bundle the flexibility of individual consumers or businesses into a portfolio of devices that are either switched on or off, depending on grid stabilization requirements. In so doing, aggregators enable smaller system users (consumers or producers) to participate indirectly in the market through the provision of flexibility services, and to receive financial benefits in return. An example of this is Direct Load Control (DLC), in which the aggregator directly controls a set of appliances within the end-user's premises [11].

Recent developments in the energy market liberalization process have created several opportunities for aggregators for the provision of market-based products and services [12]. In the Netherlands, short-term frequency deviations are balanced through the Frequency Containment Reserve (FCR) market, also known as primary reserve [13]. In the FCR market, a bidding system is applied in which parties offer a certain amount of flexible power that they can deliver whenever necessary. In return, they receive financial compensation from the Transmission System Operator (TSO) for the capacity they have offered. Given that FCR is important to system reliability, it must be provided continuously, with an availability of 100% [14]. Not meeting the promised bid results in a fine, which the DR-aggregator must pay to the TSO.

Aggregators need to construct their flexible portfolio of DR assets in such a way that it delivers the promised amount of flexibility while meeting the prerequisites of the FCR market. With a given portfolio, aggregators can choose how much flexibility they are willing to provide during the next bidding period. Determining the bid size for each bidding period is a strategic process. If the aggregator bids too low, revenue, and thus profit, can be suboptimal. On the other hand, if the aggregator bids too high, the aggregator might be unable to deliver the flexibility, and thus risks a fine. The length of the bidding period is country- and market-specific. When the bid size is determined, the aggregator is bound to deliver that amount of flexibility during the complete bidding period, in response to frequency deviations [13].

Many different technologies have the potential to operate as DR-assets, in the residential, commercial and industrial sectors. An example is the residential heat pump. Heat pumps convert electrical power into heat, used for heating households and supplying hot tap water, and thus, have a fundamental role in efficient energy use in residential buildings [15]. Heat pumps show large potential in abating CO_2 emissions, and this is accelerated by increasing shares of renewable electricity generation [16,17]. In contrast to gas-fired boilers, heat pumps are most efficient when operating at low temperatures, and are therefore considered slow response heating systems [18]. This may be a positive aspect from the perspective of switching them on or off in a DR event. Even though it is widely recognized that heat pumps can be used as flexibility assets in DR-portfolios, their flexibility is currently only rarely utilized [19].

Most research conducted in this field of study focusses on the technical performance of heat pumps in providing flexibility [20–22]. However, no literature studies were found that investigate aggregator bidding process optimization strategies. Also, the potential financial revenues resulting from offering flexibility on the Dutch FCR market using these strategies seem largely unknown. Therefore, the scientific contribution of this work lies in:

- Insights into the effects of potential aggregator bidding strategies that on the potential of FCR
- An assessment of the economic potential of domestic heat pumps to deliver flexibility on the FCR market. This potential is measured in revenue per household
- The development of a quantitative model to assess the potential of FCR, and an explanation of the model logic
- A detailed assessment of the effects of TSO-regulation on the potential for FCR

This study aims to investigate the technical and economic potential for a portfolio of aggregated residential heat pumps to provide flexibility on the Dutch FCR market. Economic potential is measured by revenue generated from providing FCR, whereas technical potential is measured in terms of bid capacity given technical constraints. In addition to the technical and economic potential, the effect of fine regulations, i.e., fines for non-availability and inadequate response, is thoroughly investigated.

Two strategies are considered that the aggregator can apply to determine the weekly bid size: the 'always available' strategy and the 'always reliable' strategy, both based on availability and/or reliability. Both strategies, as well as the concepts of reliability and availability, are explained in the method. In addition, an explanation is provided of how the model functions and how the results can be interpreted.

The paper is structured as follows. The methodology section starts with an explanation of the TSO-regulations, bidding strategies and model details. Information is then provided regarding data processing and frequency analysis, followed by a detailed explanation and overview of the model functionalities. The results section begins with the general model results. This is followed by an analysis of reliability, availability and monetary flows, and an analysis of grid frequency. After the discussion and conclusion section, the appendix provides more details regarding the TSO fine regime and household availability.

2. Method

2.1. TenneT Fine Regime and Product Specifications

DR is limited by three regulatory factors: minimum bid size, minimum bid duration and binding upward and downward bids [23]. The most important specifications for the Dutch FCR market are described in Table 1 based on a document describing the product specifications.

Table 1. FCR product specifications [24]

Specification	Description	Value & Unit
Bidding period	The length of the period over which a bid is placed. During this period, the bidding party must always be able to deliver the amount of flexibility that has been bid.	Weekly
Minimum bid size	The minimum bidding capacity. Bids with lower capacities are not processed.	1 MW
FCR full-activation time	Within this time frame, the portfolio must be fully activated (100% capacity). Due to the 5-min resolution of the data, the full activation time is not taken into account in this study.	30 s
Insensitivity range	The range of frequency deviation to which the response of the system is insensitive.	10 mHz
Full activation deviation	The maximum frequency deviation to which the system must respond with full capacity.	100 mHz

In cases where the aggregator is not available or not able to respond adequately, the aggregator will be fined by the TSO. Two types of fines are enforced in the Dutch FCR-market: fines for non-availability (NA-fines) and fines for inadequate response (IR-fines) [25]. NA-fines are imposed when the available flexible power is lower than the bid capacity. Hence, this fine can also be imposed when this flexible power is not requested by the Distribution System Operator (DSO). In contrast, IR-fines are only imposed when the aggregator does not respond adequately to a given frequency deviation. A description of both fine regimes is provided in Appendix A—TenneT TSO Fine Regime, as specified in a framework agreement concerning primary reserve [13].

2.2. Bidding Strategies

In this study, two different strategies are considered for determining how much capacity to bid on the FCR market, i.e., the 'always reliable' strategy, and the 'always available' strategy. The 'always available' strategy aims for perfect performance. It does so by determining bid size in such a way that with a given portfolio consumption, the aggregator is always able to deliver 100% of the bid capacity in both directions, at any frequency deviation. In so doing, the aggregator risks neither NA-, nor IR-fines. The 'always reliable' strategy aims to deliver 100% service reliability by choosing a bid size in which the portfolio has sufficient capacity to respond to the historical frequency deviations that are used as input for the model. This strategy does not result in any IR-fines, since given the determined bid size, the aggregator is able to respond to the historical frequency deviations. However, unlike the 'always available' strategy, this strategy will lead to NA-fines, since the aggregator is not always able to respond correctly to a 100% frequency deviation. The 'always reliable' strategy can be considered a low-risk strategy compared to the 'always available' strategy, since the chosen bid size, and therefore, the revenue, is expected to be lower.

2.3. Model Details

For this study, a quantitative model has been developed in Python (V3.5.2, Anaconda version 4.2.0, 2016,) in which historical frequency and heat pump data are used to simulate a bid process. Data from 33 households is used and scaled up to represent a portfolio of 20,000 heat pumps of 10 MW aggregated capacity with time-steps of 5 min Based on historical frequency measurements, the required amount of flexibility for every 5-min interval was determined. By simulating the switching of the heat pumps, the revenue and the fines were calculated for an iteratively increasing bid size. This process was performed for every week, leading to weekly revenue, availability percentage and reliability percentage.

2.4. Data Selection and Frequency Analysis

The data used for this study comes from Energiekoplopers, a Dutch Pilot project in Heerhugowaard aimed at assessing the potential of flexibility in residential energy systems [26]. The total dataset that was available for this study consisted of 33 households containing a heat pump. The heat pumps used were air-sourced heat pumps from the brand Inventum Ecolution Combi 50. Their minimum electrical capacity was 5W and their maximum was 500W. From these heat pumps, data presenting the power demand for the heat pump per 5 min was used for a period of 30 weeks, between 01-09-2016 and 01-05-2017. No data was available from the end of December until the beginning of February. As a result of this missing data, an 8-week gap occurs in this period, leaving 22 weeks of useful data (see Appendix B—Number of households per week).

Due to measurement errors during the pilot demonstration project, several data gaps occur in the dataset and the start date and end date between when data is available varies strongly per household. Since the bidding period as defined by TenneT is a week, the dataset was split into datafiles of one week. Data points containing extreme outliers were excluded. In addition, when more than 60 consecutive min. of heat pump data were missing, the data of that week were deleted for that heat pump. These research choices resulted in a relatively small, but reliable dataset. In order to obtain a viable portfolio size of 10 MW, the number of households was scaled up to 20,000 for every week.

2.5. Determining Availability, Reliability and NA-fines

To calculate the NA-fines for every bid size, upper and lower boundaries were determined. The upper boundary was calculated by subtracting the bid size from the maximum possible power consumption, whereas the lower boundary is calculated by adding the bid size to the minimum possible power consumption. When the power consumption exceeds those boundaries, the portfolio is not able to deliver 100% flexibility in that direction, resulting in a fine that follows the fine regime (Section 2.1) used in this study. The availability is then calculated as the number of NA-events divided by the amount of data points per week, which is 2016 5-min timestamps.

To determine reliability at a given bid size, an assessment of whether the portfolio was able to respond correctly needs to be made for every timestamp in the model. In this study, the required portfolio response is expressed in terms of Required Flexible Power (RFP). This describes the power that the portfolio should shift at each moment. A negative RFP represents a downward shift and a positive RFP represents an upwards shift to the baseline. When the portfolio is not able to deliver the RFP, a so called 'IR-event' occurs. The reliability can then be calculated as the percentage of timestamps in which no IR-event occurred.

Determining whether an IR-event occurs is a repetitive process performed in multiple steps. First, a list of available households is generated, and a household is selected from that list. Then, a household is taken from that list and the Available Flexible Power (AFP) of that household is added to the Total Available Flexible Power (TAFP), after which the household is removed from the list and the availability is updated. This process is repeated until either the list of available households is empty, or the TAFP exceeds the RFP. If the list of available households is empty before the TAFP exceeds the RFP, then an IR-event occurs. If not, then no IR-event occurs. In both cases, the model breaks out of the loop and continues to the next 5-min interval. In Figure 1, this process is illustrated.

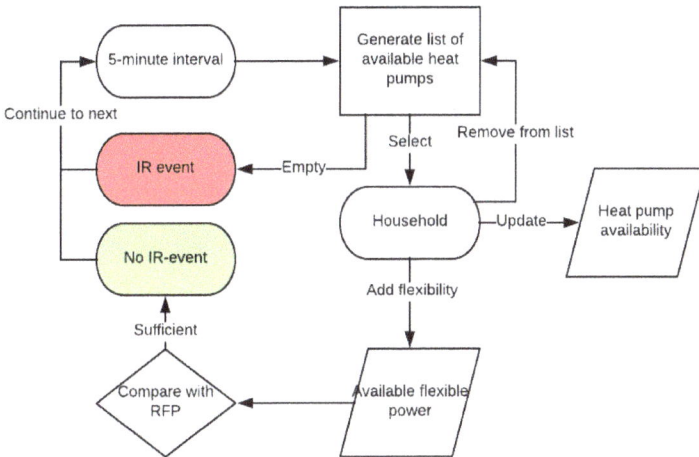

Figure 1. Switching process, aiming to determine whether an IR-event occurs in a 5-min interval.

The RFP can be calculated by dividing the frequency deviation ($F_{actual} - F_{target}$) by the Full Activation Deviation (FAD) and multiplying it with the bid size:

$$RFP(t) = Bidsize * \frac{F_{actual} - F_{target}}{FAD} \tag{1}$$

Since no more than 100% flexibility is required, the RFP cannot exceed (positively or negative) the bid size. The upper and lower boundaries of the RFP can be calculated by adding or subtracting the insensitivity range to the actual frequency (F_{actual}). In Figure 2, the portfolio activation fraction is displayed, representing the percentage of the portfolio that should be activated at any given frequency. This method is applied for every frequency measurement in the model to calculate the RFP.

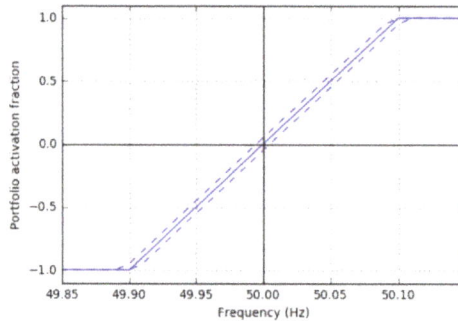

Figure 2. Portfolio activation as a response to frequency.

The main drawback of using heat pumps for DR-purposes lies in the constraints of the end-users that require a room temperature that is comfortable to live in [27]. In the model, this constraint is implemented in a simplified way by implementing a maximum switch time, thereby limiting the time that the heat pumps can be switched for. To enforce this principle, an availability-module is implemented into the model that limits the heat pump's availability for switching to a maximum of 15 min at a time. In this module, the availability status of every heat pump is stored. Heat pump availability in both directions will be stored in a data frame. The availability module works with a value for availability that is checked and updated for every 5-min interval in the model.

When the heat pump is switched, 5 min are added to the (positive) value of heat pump availability for that household at that moment. When the heat pump is not switched while it is non-active (availability < 0), 5 min are added to the availability, making it less negative. When the availability changes from −5 to 0, the heat pump will be available for switching again. Only heat pumps with a value of availability that is equal to 0 or positive can be switched. This process is illustrated in Figure 3.

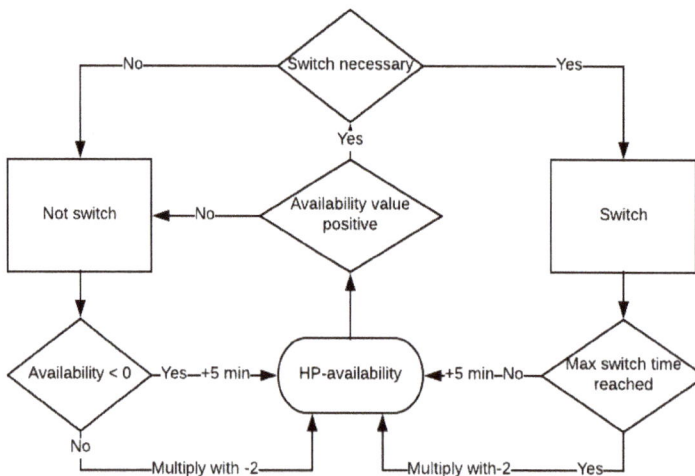

Figure 3. The process of updating availability of the heat pump, after the heat pump is switched or not.

To select heat pumps that should be switched first, heat pumps are divided in two categories:

1. Heat pumps with HP-availability > 0: these were switched in the previous timestamp but are still available. Switching these heat pumps first is most efficient.
2. Heat pumps with HP-availability = 0: these are available and were not switched in the previous timestamp. These should be switched when no category 1 heat pumps are available.

To find the heat pump that should be switched, the model first iterates over the category 1 heat pumps. If no available heat pumps exist within this group, the model will start iterating over the category 2 heat pumps. Within both groups, the algorithm looks for the heat pump that has the highest contribution of flexibility related to the RFP. It calculates the absolute difference between AFP and RFP for every heat pump. The heat pump with the highest flexibility potential will be selected as the heat pump to be switched.

2.6. Determining Bid Size and Associated/Potential Revenues

To obtain the bid size and revenue for both strategies, the model iterates over an increasing bid size, obtaining the reliability and availability for each iteration. When the reliability drops below 100%, the 'always reliable' bid size is selected as the bid size in the previous iteration. The 'always available' strategy bid size is determined in the same manner. For the main results, the bid size is increased in steps of 100 kW, starting with a minimum bid size of 100 kW. A relatively small bid size step provides high accuracy, resulting in smooth graphs and accurate results.

The revenue is based on the FCR price, expressed in €/MW/week. These prices are received from ENTSO-E (2018) and differ per week. They are based on the highest bid price in the given period. Therefore, in this study, it is assumed that the bid price equals the FCR price. In the period that is relevant for this study, prices range from €1936.77/MW/week to €3354.80/MW/week, with an average of €2559.49/MW/week. The revenue per week can be calculated by:

$$Revenue_{week(x)} = Bidsize * FCRprice \qquad (2)$$

2.7. Model Overview

In Figure 4, a visualization of the model is presented. The input for the model consists of the raw heat pump data, the raw frequency data, FCR product specifications and comfort constraints of the households. This input forms the basis for the AFP, RFP and heat pump availability, which are used for the switching process. When the switching process is repeated for every 5-min interval in the model, the availability and reliability are determined for the given bid size. The model starts with a low bid size, while iteratively increasing it until both the reliability and availability drop below 100%. The bid size at which the availability first drops below 100% was the bid size for the 'always available' strategy, whereas the bid size at which the reliability first drops below 100% was the bid size for the 'always reliable' strategy. Both revenues can be calculated based on the bid size.

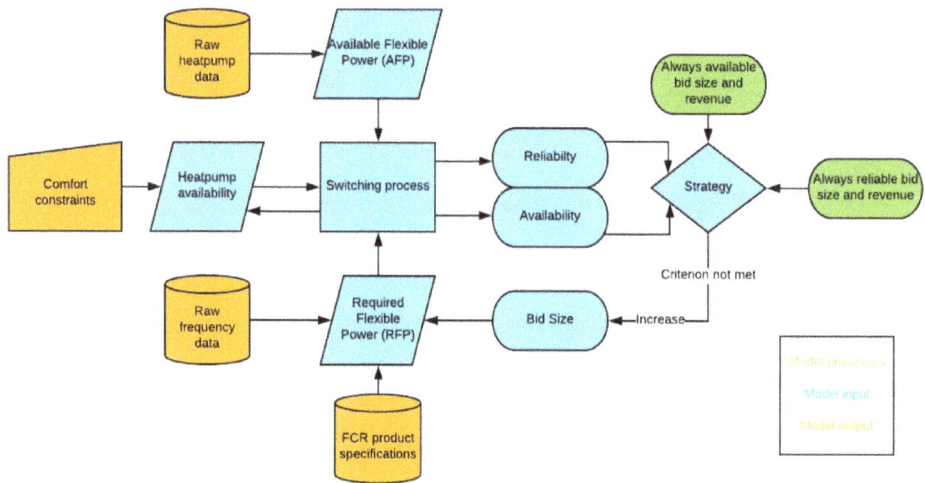

Figure 4. Model visualization.

3. Results

3.1. General Results, a Comparison between both Strategies

By using the 'always available' strategy, the aggregator successfully aims for a bid size that results in zero fines. As a result, both the availability percentage as well as the reliability are 100%. This strategy yields a lower bid size and revenue compared to the 'always reliable' strategy. With this strategy, a total of €96,114 can be earned, with an average bid size of 1.7 MW. With the 'always reliable' strategy, the total revenue is €438,318 with an average bid size of 7.9 MW. However, even though the aggregator was able to respond correctly to given frequency deviations, the low availability of only 9% that results from this strategy leads to €2,220,398 fines for non-availability. In Tables 2 and 3, an overview is presented of the main results for both strategies.

Table 2. Results for the 'always available' strategy.

	Bid Size (MW)	Revenue (€/week)	Revenue/Household (€/week/household)	Availability (%)	NA-Fines (€/week)
week01	0.3	€648,-	€0.03	100%	€-
week02	1.0	€2190,-	€0.11	100%	€-
week03	1.0	€2350,-	€0.12	100%	€-
week04	1.7	€4148,-	€0.21	100%	€-
week05	1.5	€3840,-	€0.19	100%	€-
week06	2.9	€6960,-	€0.35	100%	€-
week07	3.1	€7192,-	€0.36	100%	€-
week08	2.9	€6873,-	€0.34	100%	€-
week09	3.0	€7170,-	€0.36	100%	€-
week10	3.0	€7380,-	€0.37	100%	€-
week11	2.6	€6266,-	€0.31	100%	€-
week20	2.9	€8410,-	€0.42	100%	€-
week21	1.6	€5136,-	€0.26	100%	€-

Table 2. *Cont.*

	Bid Size (MW)	Revenue (€/week)	Revenue/Household (€/week/household)	Availability (%)	NA-Fines (€/week)
week22	1.4	€4340,-	€0.22	100%	€-
week23	1.1	€3300,-	€0.17	100%	€-
week24	1.3	€3718,-	€0.19	100%	€-
week25	2.0	€5320,-	€0.27	100%	€-
week26	0.4	€964,-	€0.05	100%	€-
week27	0.8	€1760,-	€0.09	100%	€-
week28	1.4	€2954,-	€0.15	100%	€-
week29	2.3	€4968,-	€0.25	100%	€-
week30	0.1	€227,-	€0.01	100%	€-
Average	**1.7**	**€4368,-**	**€0.22**	**100%**	**€-**
Total	**N/A**	**€96,114,-**	**€4.81**	**N/A**	**€-**

Table 3. Results for the 'always reliable' strategy.

	Bid Size (MW)	Revenue (€/week)	Revenue/Household (€/household/week)	Availability (%)	NA-fines (€/week)
week01	1.8	€3888,-	€0.19	61%	€5025,-
week02	5.7	€12,483,-	€0.62	0%	€41,921,-
week03	5.8	€13,630,-	€0.68	0%	€59,075,-
week04	8.8	€21,472,-	€1.07	0%	€120,521,-
week05	7.0	€17,920,-	€0.90	0%	€81,932,-
week06	11.2	€26,880,-	€1.34	0%	€163,223,-
week07	11.3	€26,216,-	€1.31	0%	€160,418,-
week08	11.6	€27,492,-	€1.37	0%	€170,114,-
week09	10.4	€24,856,-	€1.24	0%	€142,704,-
week10	11.2	€27,552,-	€1.38	0%	€167,152,-
week11	8.5	€20,485,-	€1.02	0%	€99,575,-
week20	11.9	€34,510,-	€1.73	0%	€216,253,-
week21	9.8	€31,458,-	€1.57	0%	€179,374,-
week22	10.1	€31,310,-	€1.57	0%	€183,240,-
week23	4.9	€14,700,-	€0.74	5%	€14,330,-
week24	7.3	€20,878,-	€1.04	0%	€90,664,-
week25	7.0	€18,620,-	€0.93	0%	€81,174,-
week26	3.7	€8917,-	€0.45	36%	€16,668,-
week27	3.1	€6820,-	€0.34	53%	€8252,-
week28	8.2	€17,302,-	€0.87	0%	€91,692,-
week29	9.8	€21,168,-	€1.06	0%	€118,667,-
week30	4.3	€9761,-	€0.49	46%	€8423,-
Average	**7.9**	**€19,923,-**	**€1.00**	**9%**	**€100,927,-**
Total	**N/A**	**€438,318,-**	**€21.92**	**N/A**	**€2,220,399,-**

3.2. Reliability and Availability, Monetary Flows and upper and lower Boundaries to Power Consumption

The upper and lower boundaries are calculated by the methods explained in Section 2.5. When the upper and lower boundaries are exceeded by the power consumption, NA-fines occur, since the portfolio is not able to deliver the capacity required according to the corresponding bid size. This does not happen, since the bid size is chosen so that no fines will occur, leading to an availability of 100% with the 'always reliable' strategy. This is illustrated in Figure 5. The reliable bid size is therefore limited by the most extreme (upper or lower) values of the power consumption.

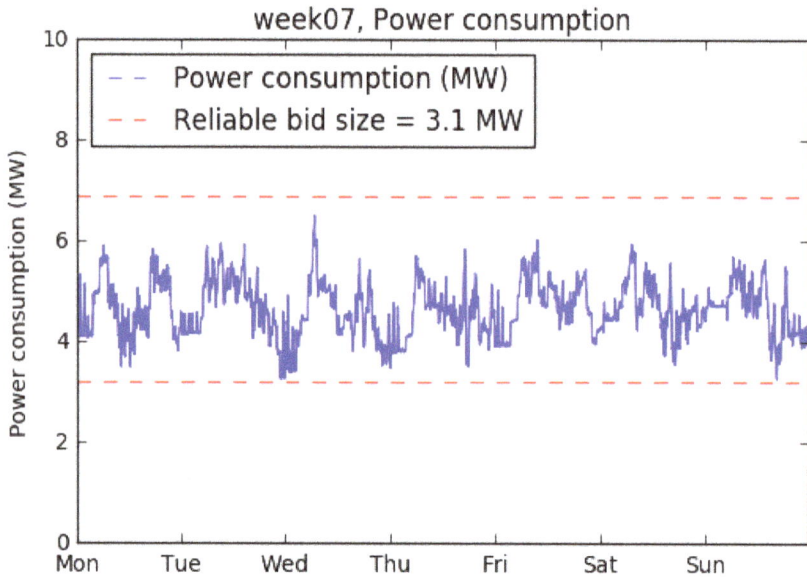

Figure 5. Power consumption (blue line) of the portfolio and boundaries (red dotted lines) at the reliable bid size of 3.1 MW.

When the bid size exceeds half the portfolio capacity (5 MW), the lower boundary will become larger than the upper boundary, making it impossible for the portfolio to remain between the boundaries and deliver the required flexibility. In these cases, the availability drops to 0%, which results in NA-fines for every measurement. Since the boundaries resulting from the 'always available' strategy are extreme, they are not displayed in Figure 5.

Portfolio-availability represents the proportion of the week in which 100% flexibility can be delivered with the portfolio, whereas the reliability represents the fraction of the week in which the portfolio responded correctly given the frequency and corresponding RFP. The major difference between portfolio availability and reliability is that the reliability is strongly influenced by the frequency and RFP, whereas the portfolio availability is solely dependent on the power consumption of the portfolio and the bid size.

Figure 6 shows a steep decrease in availability, dropping from 100% availability at a reliable bid size of 3.1 MW to 0% availability at a bid size of 5.0 MW. Availability reduction to 0% at a bid size of 5.0 MW is explained by the fact that the portfolio will not be able to deliver 100% flexibility on a symmetrical market when the bid size exceeds half the maximum capacity. Therefore, in the model, the availability is always reduced to 0% when the bid size exceeds 5.0 MW. In contrast to the availability, the reliability will not drop to 0%. Even at extremely high bid sizes, when the frequency is 50 Hz, zero flexibility is required, and the portfolio is still able to respond correctly. This frequency-dependency is the main reason that the reliability shows a less-steep decline compared to the availability. Prevalence of frequency deviation occurrence is discussed in Section 3.3.

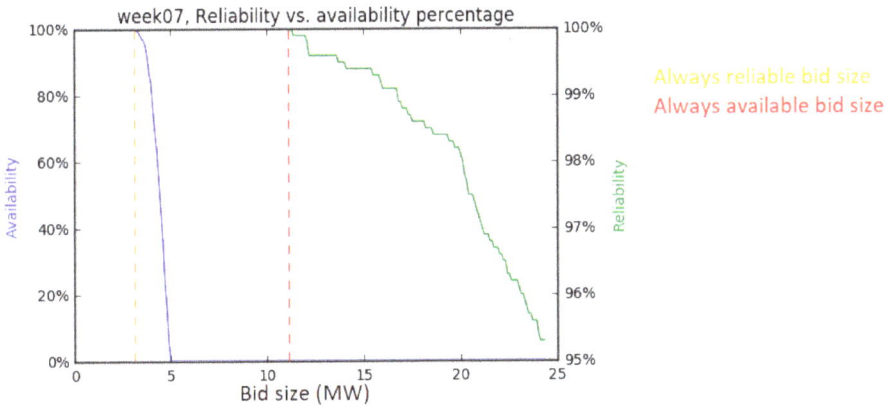

Figure 6. Availability and reliability against the bid size.

3.3. Frequency Analysis

In Table 4, the average and maximum deviation and portfolio percentage are displayed for the original dataset and for the dataset resulting from the so-called 'actual resampling method'. Using the actual resampling method has a diminishing effect on the average and maximum deviation, and thus, on the portfolio activation percentage. Given the small change in average deviation relative to the original dataset, the effect of the actual resampling method on the main results is considered minimal.

Table 4. Average and max frequency deviations.

Resampling Method	Average Deviation	Max. Deviation	Average Portfolio Activation	Max Portfolio Activation
Original	17 mHz	140 mHz	8.5%	70.0%
Actual	16 mHz	110 mHz	8.0%	55.0%

Figure 7 shows two plots, displaying the frequency distribution (top), and a distribution of the frequency deviation (bottom). In both cases, stronger deviations are less frequent compared to small deviations. This effect is visible to a larger extent in the mean dataset than in the original dataset. Frequency deviations of 0.1 Hz, in which 50% of the portfolio needs to be activated, seldom occur.

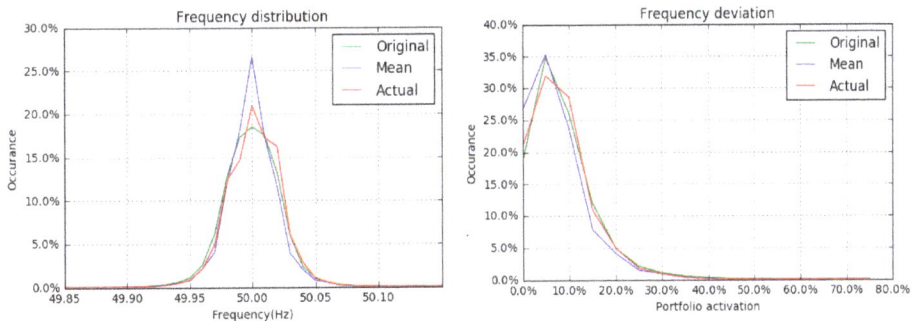

Figure 7. Distribution of frequency deviations.

4. Discussion

Results show that availability is a stronger limiting factor to bid size and revenue than reliability. To make this effect visible, an 'always reliable' strategy was implemented, in which 100% availability was not a prerequisite. With this strategy, NA-fines were calculated, but did not affect the bid size and revenue. The NA-fines were displayed to give insights into the fines that would result if the aggregator would apply this strategy. Given the high risk of fines, the 'always reliable' strategy does not seem realistic to apply in practice. However, implementing it in the model shows that it is difficult for the TSO to apply NA-fines in practice. For the 'always reliable' strategy to be implemented successfully, a perfect knowledge about frequency deviations is required. In practice, prediction algorithms might make a rough estimation of the frequency deviations, but perfect knowledge about the frequency deviations one week in advance is not feasible. Therefore, selecting a bid size with the 'always reliable' strategy is merely a theoretical concept.

An important factor when switching heating systems for DR is that the comfort of households should not be jeopardized. Ideally, the effect of heat pump switching on household temperature should be included in the model by simulating household temperature. However, the dataset lacked the information required to perform this kind of analysis. Therefore, instead of temperature boundaries, a limit was set in this model on the length of time that heat pumps could be continuously switched. This limit was set to 15 min, after which a period of non-availability was implemented. During this period, the power consumption of the heat pump follows the baseline, as it would without any interference of a third-party aggregator.

An important factor for discussion in this study is the low data quality and availability resulting in many gaps and periods with irregular values. According to the method described, these gaps and constant values were either filled or filtered out, resulting in a small but reliable sample of data. Part of the data was excluded, decreasing the amount of viable data. Eight weeks of data were missing in December and January, usually the coldest months with the highest heating potential, which might lead to a slight underestimation of the potential for FCR. To correct for the small sample size, the portfolio of households has been scaled up to mimic a larger portfolio. By doing so, data has been duplicated to generate a 10 MW portfolio. This process might influence the results of this study, since these duplication methods lead to multiple heat pumps with the same fluctuation. In practice, 20,000 heat pumps, each with a unique baseline, will generate a more stable baseline when combined. With the frequency data, these problems did not occur. Certain research design choices were implemented to provide a reliable but rather conservative estimation of the economic and technical potential of residential heat pumps in the Dutch FCR market. This was due to the implications of the data availability and quality originating from this early pilot demonstration project. The main contribution of this paper is the proposed framework and the method and logic behind the model, which can be replicated for similar studies.

Another factor that may influence the outcome of this study is the resolution of the dataset. The household data was provided on a 5-min basis, whereas the frequency data was provided on a 10-second basis. In order to reduce the complexity of the model, the 5-min resolution was used as the model resolution. The frequency data was therefore down sampled from 10 s. to 5 min by using the methods described in Section 2.4

As a consequence, short term frequency deviations (within a 5-min time framework) could not be taken into account. For this reason, the FCR specification of 30 s response time could not be taken into account either.

TSO-regulation on what is considered an IR-event or NA-event is ambiguous. Therefore, in this research, an IR-event is defined as one 5-min interval in which the aggregator was unable to respond correctly. An NA-event is defined as one 5-min interval in which the portfolio has insufficient capacity to respond to an extreme (100% portfolio activation) deviation. In addition, the model holds the assumption that an IR-event or NA-event will lead to a fine in all cases. In practice, this might not be the case, since TSOs do not have the capacity to assess and verify every IR- or NA-event and respond accurately according to the fine regime. With an increasing participation of decentralized small assets in ancillary services markets, it would require automated verification methods to support the financial settlement. To obtain more accurate results, more specific information regarding TSO-regulation is required, as well as a higher resolution and quality of the dataset.

5. Conclusions and Recommendations for Further Research

5.1. Conclusions

The main research question concerned the technical potential, and the economic potential of heat pumps to deliver ancillary services in the Dutch FCR market. The results show that both the technical and economic potential depend strongly on the bid strategy; the revenue resulting from this study is €0.22 per household per week in the 'always available' strategy, versus €1.00 per household per week in the 'always reliable' strategy. Bid sizes vary from 1.7 MW with the 'always available' strategy to 3.1 MW with the 'always reliable' strategy.

The significant difference in potential between the two strategies shows that availability is a stronger limiting factor to the potential for FCR than reliability. Table 3 shows that punishments for not being available to respond correctly to extreme frequency deviations are severe, even though these extreme frequency deviations seldom occur. It might be worthwhile reassessing the structure of the markets for ancillary services to investigate whether more flexibility could be unlocked.

Even though the results show that a considerable amount of revenue could be generated, and flexibility could be delivered, this has to be divided among 20,000 households. In order to make such a project economically feasible, marginal costs per household need to be kept extremely low. This would be challenging for any aggregator. However, the households in this model were equipped with small heat pump systems that have a peak power of only 0.5 kW. Households with larger heat pumps will be able to deliver more flexibility, thereby lowering the number of households, and thus leading to lower costs. By focusing on projects with high-capacity heat pumps, the number of households, and therefore the investment costs, can be reduced, whilst different revenue streams could be explored; for example, operating on different balancing markets or enhancing self-consumption of photovoltaic-generated electricity.

Since a strong correlation exists between outside temperature and heat pump capacity, the potential to deliver flexibility with heat pumps is strongly dependent on season. Results show that with the case study portfolio, 71% of the IR-events were IR-down events, which indicates that downward flexibility is a limiting factor in delivering FCR.

5.2. Recommendations for Further Research

In order to obtain more accurate results, a dataset of higher quality is required. With such a dataset, the effect of heat pump power consumption on room temperature can be estimated. With this effect known, households and their temperature behavior could be simulated, mimicking a real-life situation with high accuracy. In so doing, the maximum switch time, non-activity time and compensation algorithm would not be needed. An alternative solution to this approach would be to create a thermodynamic model that simulates household behavior based on insulation values and outside temperature. A sensitivity analysis could investigate the effects of the maximum switching time, non-availability factor and several FCR product specifications on the potential for FCR. The product specifications could be included in the sensitivity analysis to investigate the effect of TSO-regulation on the bidding strategies of aggregator parties. This would require cooperation between the TSO and the aggregator to clearly design the detailed regulations.

Results show that NA-fines comprise of a stronger limiting factor to the bid size and revenue compared to IR-fines. In practice, this means that the aggregator receives high fines for inability to deliver 100% flexibility, while this situation seldom occurs. Further research should aim to investigate the quality and potential of FCR by DR with a different regulation structure.

The scope of this study lies in the potential for residential heat pumps to offer flexibility on the FCR market. Further research could be performed by extending the model to operate on other markets or other technologies as well. In the Netherlands, the model could be extended to secondary or tertiary reserves, other technologies and the effect of combining different technologies on other markets. Eventually, comparisons could be made between countries and their regulations to determine how the balancing system can be optimized at the European level. The model can be used in a predictive manner, with a given portfolio, to predict in which markets the most profits can be achieved.

In this study, a portfolio is used consisting solely of residential heat pumps. In practice, given the high seasonal dependency and the fact that combining heat pumps with other DR-assets will increase the potential for FCR, it is unlikely that an aggregator will bid on the FCR market with a portfolio consisting solely of heat pumps. Future research may focus on combining the heat pumps in an integrated DR portfolio, thereby increasing the overall potential.

Author Contributions: J.P. developed the model and performed the experiments. J.P., I.L., W.S. and W.v.S. together defined the problem, formulated the research goal, revised the model, analyzed the results and wrote the paper.

Funding: This project is part of the PVProsumers4Grid Project, which received funding from the European Union's Horizon 2020 research and innovation programme under grant agreement No 764786. Furthermore, this work has received funding in the framework of the joint programming initiative ERA-Net Smart Grids Plus as part of the CESEPS project.

Conflicts of Interest: The authors declare no conflict of interest.

Appendix A. —TenneT TSO Fine Regime

To calculate the magnitude of the IR- and NA-fines, regulations that are described in a framework agreement concerning primary reserve are used [13]. In article 8, section 3.A of the framework agreement, the fine regime for NA-fines is described as follows:

'In the event of Non-Availability, supplier owes TenneT a Non-Availability Payment in proportion to the relevant Non-Availability period (which is rounded up to whole hours). The amount of the payment is calculated as follows: (10 × bid price × volume non-available power = Non-Availability payment). The bid awarded to supplier for the relevant period of the supply contract with the highest bid price is used as bid price.'

In article 9, section 1 of the framework agreement, the fine regime for IR-fines is described as follows:

'For each event where a power change (ΔP) of a technical unit is demonstrably (graph) insufficient: deduction of one 24-h period payment (= sum of the awarded bids to the supplier for the week in question), in proportion with the primary reserve which is reserved for the technical unit in question (from allocation message of supplier). For every supply contract, the compensation for inadequate response by supplier to TenneT is maximized at 3 times the sum of the awarded bids to supplier for the week in question.'

Appendix B. —Number of Households per Week

Week Number	From	To	Households	Without Gaps	Deleted
1	3-10-2016	10-10-2016	22	21	1
2	10-10-2016	17-10-2016	20	20	0
3	17-10-2016	24-10-2016	22	19	3
4	24-10-2016	31-10-2016	23	23	0
5	31-10-2016	7-11-2016	19	18	1
6	7-11-2016	14-11-2016	19	19	0
7	14-11-2016	21-11-2016	24	24	0
8	21-11-2016	28-11-2016	23	22	1
9	28-11-2016	5-12-2016	21	21	0
10	5-12-2016	12-12-2016	22	19	3
11	12-12-2016	19-12-2016	21	20	1
12	19-12-2016	26-12-2016	21	0	21
13	26-12-2016	2-1-2017	0	0	0
14	2-1-2017	9-1-2017	0	0	0
15	9-1-2017	16-1-2017	0	0	0
16	16-1-2017	23-1-2017	0	0	0
17	23-1-2017	30-1-2017	0	0	0
18	30-1-2017	6-2-2017	0	0	0
19	6-2-2017	13-2-2017	13	0	13
20	13-2-2017	20-2-2017	16	16	0
21	20-2-2017	27-2-2017	16	16	0
22	27-2-2017	6-3-2017	15	13	2
23	6-3-2017	13-3-2017	19	17	2
24	13-3-2017	20-3-2017	20	19	1
25	20-3-2017	27-3-2017	19	16	3
26	27-3-2017	3-4-2017	19	19	0
27	3-4-2017	10-4-2017	22	17	5
28	10-4-2017	17-4-2017	22	20	2
29	17-4-2017	24-4-2017	23	22	1
30	24-4-2017	1-5-2017	19	19	0
Total			480	420	60

References

1. Moriarty, P.; Honnery, D. What is the global potential for renewable energy? *Renew. Sustain. Energy Rev.* **2012**, *16*, 244–252. [CrossRef]
2. Weitemeyer, S.; Kleinhans, D.; ThomasVogt, A.C. Integration of renewable energy sources in future power systems: The role of storage. *Renew. Energy* **2015**, *75*, 14–20. [CrossRef]
3. Marzband, M.; Fouladfar, M.H.; Akorede, M.F.; Lightbody, G.; Pouresmaeil, E. Framework for smart transactive energy in home-microgrids considering coalition formation and demand side. *Sustain. Cities Soc.* **2018**, *40*, 136–154. [CrossRef]

4. Mousa, M.; Azarinejadian, F.; Savaghebi, M.; Pouresmaeil, E.; Guerrero, J.M.; Lightbody, G. Smart transactive energy framework in grid-connected multiple home microgrids under independent and coalition. *Renew. Energy* **2018**, *126*, 95–106. [CrossRef]

5. Aghaei, J.; Alizadeh, M.I. Demand response in smart electricity grids equipped with renewable energy sources: A. review. *Renew. Sustain. Energy Rev.* **2013**, *18*, 64–72. [CrossRef]

6. Lampropoulos, I.; Kling, W.L.; Ribeiro, P.F.; van den Berg, J. History of Demand Side Management and Classification of Demand Response Control Schemes. In Proceedings of the 2013 IEEE PES General Meeting, Vancouver, BC, Canada, 21–25 July 2013.

7. Eurelectric. Everything You Always Wanted to Know about Demand Response. 2015. Available online: https://www3.eurelectric.org/media/176935/demand-response-brochure-11-05-final-lr-2015-2501-0002-01-e.pdf (accessed on 12 February 2018).

8. Lampropoulos, I.; Vanalme, G.; Kling, W.L. A Methodology for Modelling the Behaviour of Electricity Prosumers within the Smart Grid. In Proceedings of the IEEE PES Conference on Innovative Smart Grid Technologies Europe, Gothenburg, Sweden, 11–13 October 2010.

9. Marzband, M.; Javadi, M.; Pourmousavi, S.A.; Lightbody, G. An advance retail electricity market for active distribution systems and home microgrid interoperability based on game theory. *Electr. Power Syst. Res.* **2018**, *157*, 187–199. [CrossRef]

10. Wang, Q.; Zhang, C.; Ding, Y.; Xydis, G.; Wang, J.; Østergaard, J. Review of real-time electricity markets for integrating distributed energy resources and demand response. *Appl. Energy* **2015**, *138*, 695–706. [CrossRef]

11. Paterakis, N.G.; Erdinç, O.; Catalão, J.P. An overview of demand response: key-elements and international experience. *Renew. Sustain. Energy Rev.* **2017**, *69*, 871–891. [CrossRef]

12. Lampropoulos, I.; van den Broek, M.; van der Hoofd, E.; Hommes, K.; van Sark, W. A system perspective to the deployment of flexibility through aggregator companies in the Netherlands. *Energy Policy* **2018**, *118*, 534–551. [CrossRef]

13. TenneT. *FAQ Primary Reserve (FCR) Report*; TenneT: Arnhem, The Netherlands, 2017.

14. Thien, T.; Schweer, D.; vom Stein, D.; Moser, A.; Uwe Sauer, D. Real-world operating strategy and sensitivity analysis of frequency containment reserve provision with battery energy storage systems in the German market. *J. Energy Storage* **2017**, *13*, 143–163. [CrossRef]

15. Alessio, G.; Carli, M.D.; Zarrella, A.; Bella, A.D. Efficiency in heating operation of low-temperature radiant systems working under dynamic conditions in different kinds of buildings. *Appl. Sci.* **2018**, *8*, 2399. [CrossRef]

16. Litjens, G.B.M.A.; Worrell, E.; Sark, W.G.J.H.M.V. Energy & Buildings Lowering greenhouse gas emissions in the built environment by combining ground source heat pumps, photovoltaics and battery storage. *Energy Build.* **2018**, *180*, 51–71.

17. Ploskić, A.; Wang, Q.; Sadrizadeh, S. Mapping relevant parameters for efficient operation of low-temperature heating systems in Nordic. *Appl. Sci.* **2018**, *8*, 1973. [CrossRef]

18. Li, X.; Wu, W.; Zhang, X.; Shi, W.; Wang, B. Energy saving potential of low temperature hot water system based on air source absorption heat pump. *Appl. Therm. Eng.* **2012**, *48*, 317–324. [CrossRef]

19. Fischer, D.; Wolf, T.; Triebel, M. Flexibility of Heat Pump Pools: The Use of SG-Ready from an Aggregator's Perspective. In Proceedings of the 12th IEA Heat Pump Conference 2017, Rotterdam, The Netherlands, 15–18 May 2017.

20. Bhatarai, B.P.; Bak-Jensen, B.; Pilai, J.R.; Maier, M. Demand flexibility from residential heat pump. In Proceedings of the 2014 IEEE PES General Meeting | Conference & Exposition, National Harbor, MD, USA, 27–31 July 2014.

21. Hong, J.; Johnstone, C.; Torriti, J.; Leach, M. Discrete demand side control performance under dynamic building simulation: A heat pump application. *Renew. Energy* **2012**, *39*, 85–95. [CrossRef]

22. Fischer, D.; Madani, H. On heat pumps in smart grids: A review. *Renew. Sustain. Energy Rev.* **2017**, *70*, 342–357. [CrossRef]

23. Koliou, E.; Eid, C.; Chaves-Avila, J.; Hakvoort, R. Demand response in liberalized electricity markets: Analysis of aggregated load participation in the German balancing mechanism. *Energy* **2014**, *71*, 245–254. [CrossRef]

24. TenneT. Productspecificatie FCR. 2015. Available online: https://www.tennet.eu/fileadmin/user_upload/Company/News/Dutch/2016/productspecificatie_FCR_-_Europese_System_Operation_Guideline.pdf (accessed on 16 February 2018).

25. Energie Nederland. Response to the ENTSO-E Consultation on "FCR Cooperation" Potential Market Design Evolutions. Available online: http://www.energie-nederland.nl/app/uploads/2017/02/E-NL-response-FCR-consultation-final.pdf (accessed on 16 December 2018).

26. Energiekoplopers. Eindrapport van EnergieKoplopers Fase 1 (2015–2016). Available online: https://www.energiekoplopers.nl/ (accessed on 14 November 2017).

27. Parkinson, S.; Wang, D.; Crawford, C.; Djilali, N. Comfort-constrained distributed heat pump management. *Energy Procedia* **2011**, *12*, 849–855. [CrossRef]

applied sciences

MDPI

Article

Resonance Instability of Photovoltaic E-Bike Charging Stations: Control Parameters Analysis, Modeling and Experiment

Ziqian Zhang [1,*], Cihan Gercek [2], Herwig Renner [1], Angèle Reinders [2,3] and Lothar Fickert [1]

[1] Institute of Electrical Power Systems, Graz University of Technology, Inffeldgasse 18/I, 8010 Graz, Austria; herwig.renner@tugraz.at (H.R.); lothar.fickert@tugraz.at (L.F.)

[2] Department of Design, Production and Management, Faculty of Engineering Technology, University of Twente, P.O. Box 217, 7500 AE Enschede, The Netherlands; c.gercek@utwente.nl (C.G.); a.h.m.e.reinders@utwente.nl or a.h.m.e.reinders@tue.nl (A.R.)

[3] Energy Technology Group at Mechanical Engineering, Eindhoven University of Technology, P.O. Box 513, 5600 MB Eindhoven, The Netherlands

* Correspondence: ziqian.zhang@tugraz.at; Tel.: +43-316-873-7551; Fax: +43-316-873-7553

Received: 1 December 2018; Accepted: 6 January 2019; Published: 11 January 2019

Abstract: This article presents an in-situ comparative analysis and power quality tests of a newly developed photovoltaic charging system for e-bikes. The various control methods of the inverter are modeled and a single-phase grid-connected inverter is tested under different conditions. Models are constituted for two current control methods; the proportional resonance and the synchronous rotating frames. In order to determine the influence of the control parameters, the system is analyzed analytically in the time domain as well as in the frequency domain by simulation. The tests indicated the resonance instability of the photovoltaic inverter. The passivity impedance-based stability criterion is applied in order to analyze the phenomenon of resonance instability. In conclusion, the phase-locked loop (PLL) bandwidth and control parameters of the current loop have a major effect on the output admittance of the inverter, which should be adjusted to make the system stable.

Keywords: power quality; grid-connected inverter; photovoltaic; solar mobility; solar charging; resonance instability

---

1. Introduction

Transportation takes up a lot of space on the overall energy spending of the population. The future transmission capacity of the grid might be limited because of the increasing electricity demand, therefore electric mobility of the future would aim to be less demanding for the grid [1]. Due to the optimization of the space utilization of rooftop photovoltaic systems and charging stations, photovoltaic (PV) electric vehicle (EV) charging stations [2,3] and photovoltaic electric bicycle charging stations [4,5] have received widespread attention [6]. Since photovoltaic charging stations require sufficient sunshine time and parking space, they are generally far from buildings [7]. Therefore, the photovoltaic charging station is often at the end of the low-voltage power grid of the building, and the grid impedance has a great influence on the stable operation of the photovoltaic charging station [8]. To improve the performance of new forms of EV charging such as by solar energy, it would be interesting to evaluate and quantify the power quality of charging stations.

The common stability analysis methods of grid-connected inverters are the state space method [9] and the impedance analysis method [10]. When applying the state space method for stability analysis, all parameters of the inverter and the grid must be acquired to establish the state space model [11]. Due to the curse of dimensionality problem [12], the state space method is very complicated for the analysis process of a large-scale grid-connected inverter-grid system.

Therefore, the impedance analysis method is proposed [13]. This method considers the inverter and grid system as two independent subsystems and establishes an impedance model separately or directly measures the impedance by the frequency sweeping method. Then, through the Nyquist criterion, whether the inverter-grid system is stable or not can be judged. In this article, the passivity impedance-based stability criterion [14] and the impedance analysis method are combined to analyze the undesired resonance or harmonic phenomenon, which is called resonance instability [15,16].

In Section 2, the power quality of a photovoltaic (PV) charging station for e-bikes at the campus of the University of Twente (UT) [17] is measured and analyzed. This grid-connected PV charging station has been developed in house by UT using commercially available components (275 W$_p$ of Canadian Solar® PV modules and 2.0 kW inverter SOLAX Power® X1 Series) and was installed by the end of 2017.

The resonance instability due to the grid impedance is found in the experimental results. The phenomenon of resonance instability, its principles, and its stability criteria are discussed. In Section 3, two common single-phase photovoltaic inverter control methods are presented as models. The modeling considers the influence of the phase-locked loop (PLL) on the inverter impedance. In Section 4, the model that is presented in Section 3 is validated through simulation. In the simulation, the phenomenon of resonance instability in the experiment is reproduced. In Section 5, the influences of various parameters on the stability of the inverter in each frequency band are analyzed and discussed. In addition, the improvement methods are presented and validated through simulation. The paper ends with conclusions in Section 6.

2. Experimental Study

2.1. Experimental Set-Up

The PV charging station for e-bikes is located in the bicycle parking lot of a central building in the UT Campus, enabling the public to charge their e-bikes with solar energy. Figure 1 shows the PV charging station with six dark mono-crystalline silicon PV modules of Canadian Solar® ALL-BLACK CS6K-275 (marked with red dotted squares at the upper part of Figure 1), and a single-phase grid-connected inverter with a rated power of 2 kW (at left part of Figure 1). For further details on the PV modules and the inverter, please refer to References [18,19], respectively. One can also see in Figure 1, the flexible CIGS mini modules in the front wheel PV of the solar e-bikes.

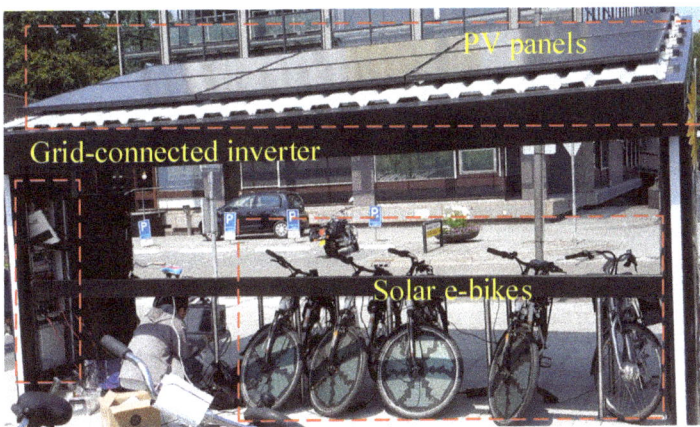

Figure 1. The PV (photovoltaic)charging station with 6 PV modules (top), a grid-connected inverter (at left column), and 4 solar e-bikes that are being charged.

The single-line diagram of the photovoltaic charging station is illustrated in Figure 2. As shown in Figure 2, six photovoltaic panels are connected to the DC side of the grid-connected inverter. The AC side of the inverter is connected to the public at the point of common coupling (PCC), so that the charging may be effectuated from the grid when there is no solar input. The PCC is located on the secondary side of the transformer of the building in Figure 1. The electrical distance between the inverter (point b) and PCC is relatively long, and there are also many non-linear loads that are switched at any time in the building. These together make the grid-connected inverter connected to a time-varying weak grid.

In order to analyze the power quality of the photovoltaic charging station, current and voltage sensors were installed in point a (DC bus), b (inverter output), c (local load) and d (grid) of Figure 2.

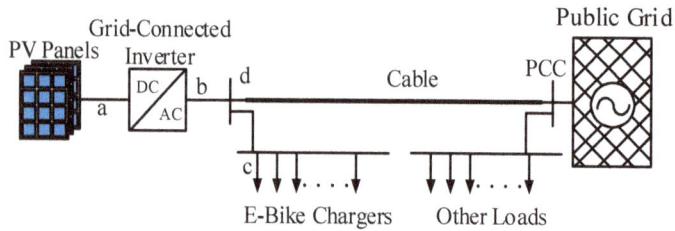

Figure 2. The single-line diagram of the PV Charging Station.

2.2. Experimental Results

In one test case, when the inverter output power and the power of local load (point c in Figure 2) did not change significantly, the total harmonic distortion (THD) of the inverter output current abnormally rose, as shown in Figure 3c.

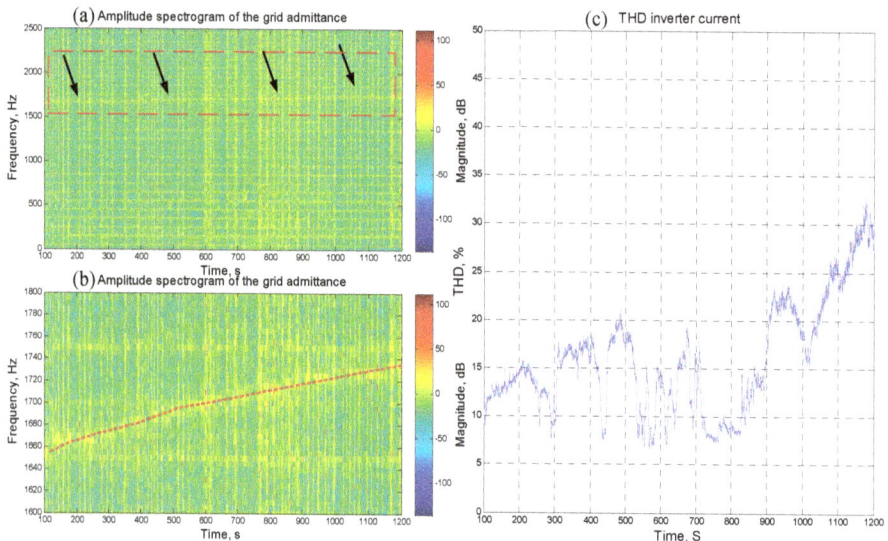

Figure 3. The measurement data for the Photovoltaic Charging Station: (**a**) grid admittance amplitude spectrogram; (**b**) partial enlargement of (**a**); (**c**) THD (total harmonic distortion) of inverter output current.

By using the passive line impedance estimation method [20,21], the grid admittance amplitude spectrogram is obtained from the grid voltage and current information, as shown in Figure 3a. Its partial enlargement is shown in Figure 3b. In Figure 3a,b, the abscissa represents time, the ordinate represents frequency, and the color in the figure represents the amplitude (dB) of the grid admittance, with warm colors representing larger amplitudes and cool colors representing smaller amplitudes.

As shown in the red dotted square, indicated by the black arrow in Figure 3a, there was a resonance component of grid admittance with about a 10 dB amplitude rising from 1650 Hz to 1730 Hz. It passed through 1730 Hz from below at around 1000 s (see the partial enlargement (Figure 3b) for a clearer view). The THD of the inverter output current also started to increase around t = 1000 s (Figure 3c). This may be caused by a change in the operating state of other loads in the grid.

The undesired increasing of the THD of the inverter output current can lead to increased copper losses and a shortened life of the transformer, and it can also cause relay protection equipment to malfunction. If the THD continues to increase, it will also cause the inverter to lose stability and stop operation.

2.3. Passivity Impedance-Based Stability Criterion

Similar phenomena were also described in Reference [22], and called "resonance instability" or "harmonic instability". This can be explained by the impedance-based stability criterion [13] and passivity theory [14]. Considering the nonlinearity of an inverter-grid system, the linearization of the equivalent circuit is based on the small signal modeling, as illustrated in Figure 4.

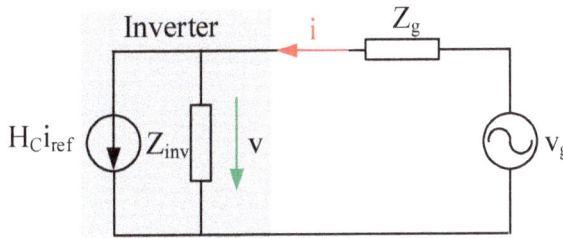

Figure 4. The equivalent circuit diagram of the inverter-grid system.

Most grid-connected inverters use the current tracking mode, so the inverter can be thought of as an ideal current source ($H_c i_{ref}$) and an inverter output impedance (Z_{inv}) connected in parallel, as shown in Figure 4, in the grey box.

The expression of the inverter output current is expressed in Equation (1).

$$i = H_C(s)i_{ref} + Y_{inv}(s)v \tag{1}$$

where i is the inverter output current; $H_c(s)$ is the transfer function of the current control loop of the inverter; i_{ref} is the reference current; $Y_{inv}(s)$ is the inverter output admittance and v is the inverter output voltage.

The complete inverter output current expression of the inverter-grid system is obtained in Equation (2), by Equation (1) and Figure 4,

$$i = \underbrace{\frac{H_C(s)}{1 + Y_{inv}(s)Z_g(s)}}_{H(s)} i_{ref} + \underbrace{\frac{Y_{inv}(s)}{1 + Y_{inv}(s)Z_g(s)}}_{Y(s)} v \tag{2}$$

$Z_g(s)$ represents the grid impedance; vg is the grid voltage, H(s) and Y(s) are the closed-loop transfer function and input admittance of the inverter-grid system, separately.

The stability of Y(s) may be analyzed by the Nyquist criterion with the open-loop transfer function $Y_{OL}(s) = Y_{inv}(s)Z_g(s)$. When there is a resonance point in the grid impedance with a small damping coefficient, the grid impedance will intersect the output impedance of the inverter near the resonance frequency ω_r with a large peak. Additionally, the phase angle of grid impedance will abruptly change from π to $-\pi$ at ω_r. When the phase angle of the inverter output admittance meets $-\pi \geq \arg[Yinv(j\omega)] \geq \pi$, then the phase angle of the open-loop transfer function $\arg[Y_{OL}(j\omega)]$ will certainly cross over $\pm 2\pi$ near the resonance frequency ω_r. Thus, Y(s) will be unstable in this situation. Conversely, if the phase angle of the inverter output admittance meets $-\pi \leq \arg[Y_{inv}(j\omega)] \leq \pi$, which can also be called passivity, Y(s) can be stable.

The passivity of a transfer function Y(s) is defined as [14]

1. Y(s) is stable and;
2. $-\pi \leq \arg[Y(j\omega)] \leq \pi$ and can also be expressed equivalently as;
3. $\text{Re}\{Y(j\omega)\} \geq 0, \forall \omega \in (0, fL]$. which means that the transfer function is passive in the interval $(0, fL]$.

Y(s) is stable since the Nyquist curve of Y(s) cannot encircle $(-1, j0)$ if it is located at the right half s plane, of which the real part is ≥ 0 in the interval $(0, fL]$.

To ensure that Y(s) is passive, $Y_{inv}(s)$ and $Z_g(s)$ need to be proven to be passive. We assume that both $Y_{inv}(s)$ and $Z_g(s)$ are passive, as shown in Equation (3).

$$\begin{cases} \Re\{Y_{inv}(j\omega)\} \geq 0 \\ \Re\{Z_g(j\omega)\} \geq 0 \end{cases}, \forall \omega \in (0, f] \tag{3}$$

By bringing Equation (3) into the Y(s) of Equation (2) to get the expression of the real part of Y(s), it is shown that Y(s) is also passive (Equation (4)).

$$\begin{aligned} \Re\{Y(j\omega)\} &= \Re\left\{ \frac{Y_{inv}(j\omega)\left(1 + \overline{Y}_{inv}(j\omega)\overline{Z}_g(j\omega)\right)}{|1 + Y_{inv}(j\omega)Z_g(j\omega)|^2} \right\} \\ &= \Re\left\{ \frac{Y_{inv}(j\omega) + Y_{inv}(j\omega)\overline{Y}_{inv}(j\omega)\overline{Z}_g(j\omega)}{|1 + Y_{inv}(j\omega)Z_g(j\omega)|^2} \right\} \\ &= \frac{\Re\{Y_{inv}(j\omega)\} + |Y_{inv}(j\omega)|^2\Re\{Z_g(j\omega)\}}{|1 + Y_{inv}(j\omega)Z_g(j\omega)|^2} \geq 0, \forall \omega \in (0, f_L] \end{aligned} \tag{4}$$

The current loop control transfer function $H_c(s)$ of a well-functioning grid-connected inverter is passive because its poles are located in the left half plane. Due to the same principle, the closed-loop transfer function of the inverter-grid system H(s) is also passive if Y(s) is already passive.

The power grid is composed of passive components such as an inductor, capacitor, and resistor, so the grid impedance $Z_g(s)$ is always stable and passive [14]. In this way, the stability of the inverter-grid system is determined by the passivity degree of the inverter output admittance Yinv(s). This criterion is expressed in Equation (5).

$$\text{Re}\{Y_{inv}(j\omega)\} \geq 0, \forall \omega \in (0, f_L] \tag{5}$$

When Equation (5) is met, the inverter-grid system is stable if the resonance frequency of grid impedance is located in the interval (0,fL]. In Equation (5), when $\text{Re}\{Y_{inv}(j\omega)\}$ is equal to 0, the inverter-grid system will be in a critical stable state. The inverter output admittance can be derived from the modeling of the inverter.

3. Modeling the Inverter in the Frequency-Domain

The research object of this article is a single-phase grid-connected inverter. Commonly used single-phase grid-connected inverter current control methods are

- proportional resonance (PR) based current control
- synchronous rotating frame (dq-transformation) based current control

3.1. PR-Based Current Control Method

The block diagram of a typical PR based current control single-phase inverter [23] is illustrated in Figure 5. The DC bus supplies power to the full bridge circuit with the voltage vdc. The output voltage of the full bridge circuit is u, i is the inverter output current. The full bridge circuit connects to the grid (v) through an inductive filter (L). The single-phase inverter obtains the phase angle of the grid voltage through a phase-locked loop in order to generate reference current i^* with the output from the DC voltage regulator i*m. The current loop controller is a quasi-PR regulator, which may robustly track the sinusoidal waveform without a static error, its transfer function is expressed in Equation (6).

$$G_{PR}(s) = K_{P-PR} + \frac{2K_{P-PR}\omega_c}{s^2 + 2\omega_c + \omega_0^2} \tag{6}$$

where $K_{P\text{-}PR}$ is the proportional coefficient; $K_{R\text{-}PR}$ is the resonance coefficient; ω_c is the control bandwidth; ω_0 is the resonant frequency.

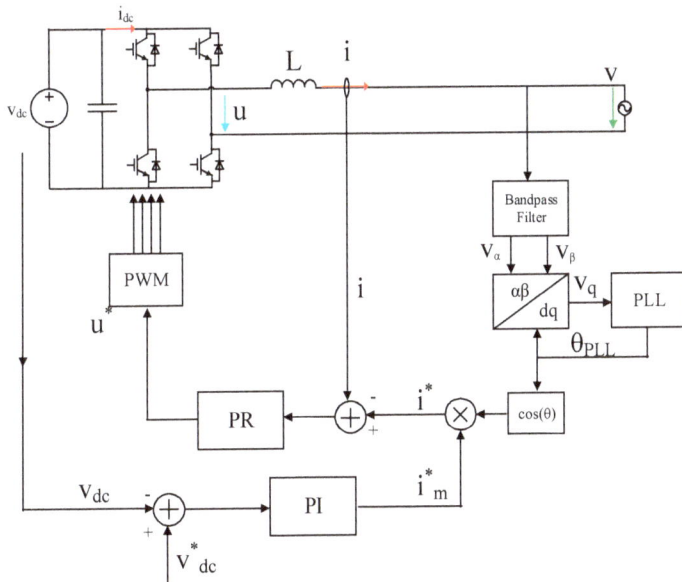

Figure 5. The block diagram of a typical PR (proportional resonance) based current control inverter.

The output of the regulator is modulation the signal u*. The transfer function of the modulation signal u* to the output voltage of the full bridge circuit u is KPWM. The dynamics on the DC side are neglected here due to the large time constant on the DC side. The small signal model of the system is illustrated in Figure 6.

In Figure 6, the grey box groups the hardware system of the inverter, the rest represents the controller system. $Y_0(s)$ is the output admittance of the inverter hardware, $G_{id}(s)$ is the transfer function from the output voltage of the full bridge circuit u to output current i. $T_{PLL}(s)$ is the transfer function from the grid side voltage v to the normalized current reference. The controller transfer function $G_{PR}(s)$ is expressed in Equation (6), $G_d(s)$ is the transfer function for the time delay in the digital control. The delay time consists of the calculation time in the controller and PWM modulation time. In this case, it is 1.5 times that of the sampling period [24,25].

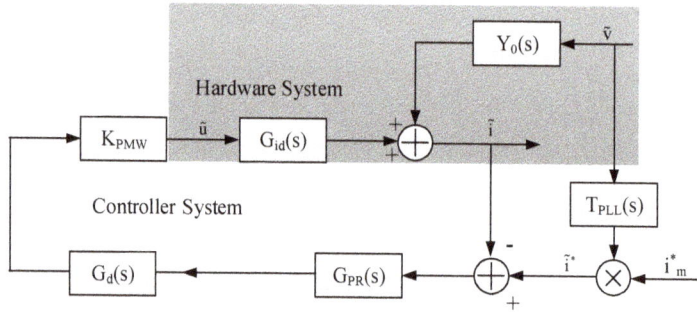

Figure 6. The small signal model of a typical PR based current control inverter.

The output admittance of the inverter hardware Y_0 can be derived by assuming the inverter output is zero; G_{id} can be derived by assuming the grid voltage is zero, as expressed in Equation (7).

$$\begin{cases} Y_0(s) = -\frac{1}{Ls} \\ G_{id}(s) = \frac{1}{Ls} \end{cases} \tag{7}$$

Without considering the effect of the PLL on the inverter, according to Figure 6, the transfer function from reference current i* to inverter output current i can be obtained, as expressed in Equation (8).

$$i = \frac{G_{PR}(s)G_d(s)K_{PWM}}{G_{PR}(s)G_d(s)K_{PWM} + Ls}i^* - \frac{1}{G_{PR}(s)G_d(s)K_{PWM} + Ls}v \tag{8}$$

As shown in Figures 5 and 6, the dynamics of the PLL will affect the performance of the inverter control, so the PLL also needs to be considered in the modeling of the inverter. The block diagram of a typical single-phase synchronous rotation frame (SRF) PLL is shown in Figure 7.

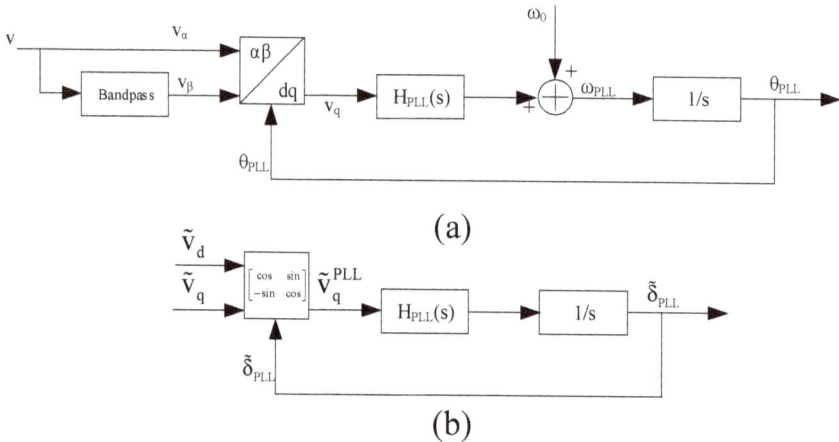

(a)

(b)

Figure 7. The block diagram of a typical SRF (synchronous rotation frame) PLL. (**a**) Block diagram of a typical single-phase SRF PLL; (**b**) Small signal model of SRF PLL.

The operation principle of a single-phase SRF PLL is that by a bandpass-filter, two orthogonal grid voltages v_α and v_β will be obtained. Then the *q*-axis component of the grid voltage v_q can be calculated by performing a dq transformation on the grid voltage. Let vq become 0 through the closed loop control. When $v_q = 0$, the output of the regulator ω_{PLL} should be the angular velocity of the grid voltage ω_g, and its integral value θ_{PLL} should be the phase angle of the grid voltage θ_g.

In the steady state case, θ_{PLL} is equal to θ_g and the dq components of the voltage v_{dq}^{PLL} calculated by the PLL is equal to the dq components of the grid voltage vdq. In the case of non-steady state behavior (by small-signal perturbation), due to the influence of the control loop in PLL, the phase angle obtained by the PLL θ_{PLL} is no longer equal to the phase angle of the grid voltage θ_g, so the calculated dq component of the grid voltage v_{dq}^{PLL} in the controller system does not represent the exact grid situation. We then set the difference between θ_{PLL} and θg as δ_{PLL}. The transfer function from the grid voltage disturbance to the voltage disturbance in the controller is expressed in Equations (9) and (10).

$$\tilde{v}_{dq}^{PLL} = T_{\delta_{PLL}}\tilde{v}_{dq} \tag{9}$$

$$T_{\delta_{PLL}} = \begin{bmatrix} \cos(\delta_{PLL}) & \sin(\delta_{PLL}) \\ -\sin(\delta_{PLL}) & \cos(\delta_{PLL}) \end{bmatrix} \tag{10}$$

where the variable with the superscript PLL represents the dq component in the controller and the variable without the superscript represents the exact dq component.

Adding a small-signal perturbation to Equations (9) and (10) results in Equation (11).

$$\begin{bmatrix} v_d^{PLL} + \tilde{v}_d^{PLL} \\ v_q^{PLL} + \tilde{v}_q^{PLL} \end{bmatrix} = \begin{bmatrix} \cos\left(\tilde{\delta}_{PLL}\right) & \sin\left(\tilde{\delta}_{PLL}\right) \\ -\sin\left(\tilde{\delta}_{PLL}\right) & \cos\left(\tilde{\delta}_{PLL}\right) \end{bmatrix} \cdot \begin{bmatrix} v_d + \tilde{v}_d \\ v_q + \tilde{v}_q \end{bmatrix} \tag{11}$$

When the phase angle perturbation is negligible, Equation (12) can be derived by Equation (11).

$$\begin{bmatrix} v_d^{PLL} + \tilde{v}_d^{PLL} \\ v_q^{PLL} + \tilde{v}_q^{PLL} \end{bmatrix} \approx \begin{bmatrix} 1 & \tilde{\delta}_{PLL} \\ -\tilde{\delta}_{PLL} & 1 \end{bmatrix} \cdot \begin{bmatrix} v_d + \tilde{v}_d \\ v_q + \tilde{v}_q \end{bmatrix} \tag{12}$$

By neglecting the steady state values in Equation (12), Equation (13) is obtained.

$$\begin{bmatrix} \tilde{v}_d^{PLL} \\ \tilde{v}_q^{PLL} \end{bmatrix} \approx \begin{bmatrix} \tilde{\delta}_{PLL}v_q + \tilde{v}_d \\ -\tilde{\delta}_{PLL}v_d + \tilde{v}_q \end{bmatrix} \tag{13}$$

According to Figure 7b, the output of the PLL can be derived as Equation (14).

$$\tilde{\delta}_{PLL} = \tilde{v}_q^{PLL}H_{PLL}(s)\frac{1}{s} \tag{14}$$

By merging Equations (13) and (14), the transfer function of the PLL can be written as in Equation (15).

$$\frac{\tilde{\delta}_{PLL}}{\tilde{v}_q} = G_{PLL}(s) = \frac{H_{PLL}(s)}{H_{PLL}(s)v_d + s} \tag{15}$$

According to the block diagram (Figure 6) and considering the orthogonality of vα and vβ [26], the transfer function from the grid side voltage to the current reference can be derived, as expressed in Equation (16).

$$\frac{\tilde{i}^*}{\tilde{v}_\alpha} = \frac{\tilde{i}^*}{\tilde{v}} = i_m^* T_{PLL}(s) = \frac{i_m^*}{2}\frac{H_{PLL}(s - j\omega_0)}{H_{PLL}(s - j\omega_0)v_d + (s - j\omega_0)} \tag{16}$$

By applying Equation (16) into Equation (8), the relationship from the grid voltage to the inverter output current by considering the influence of PLL and the inverter output admittance will be obtained as shown in Equations (17) and (18), respectively.

$$i = \frac{i_m^* T_{PLL}(s)G_{PR}(s)G_d(s)K_{PWM} - 1}{G_{PR}(s)G_d(s)K_{PWM} + Ls}v \tag{17}$$

$$Y_{inv}(s) = \frac{i_m^* T_{PLL}(s)G_{PR}(s)G_d(s)K_{PWM} - 1}{G_{PR}(s)G_d(s)K_{PWM} + Ls} \tag{18}$$

3.2. Synchronous Rotating Frame Based Current Control Method

The block diagram of a typical dq-transformation based current control single-phase inverter [23] is illustrated in Figure 8. Its hardware topology is the same as that of the PR-based current control inverter. The dq components of the inverter output current i_{dq} are obtained by dq transformation. Then the dq components of the current are controlled separately. The reference value of the dq components of the current i^*_{dq} is obtained by a DC voltage regulator or/and calculated by the active/reactive power requirements. The output of the current regulator is converted into a reference of the inverter output voltage u^*_{dq} after decoupling and inverse dq transformation.

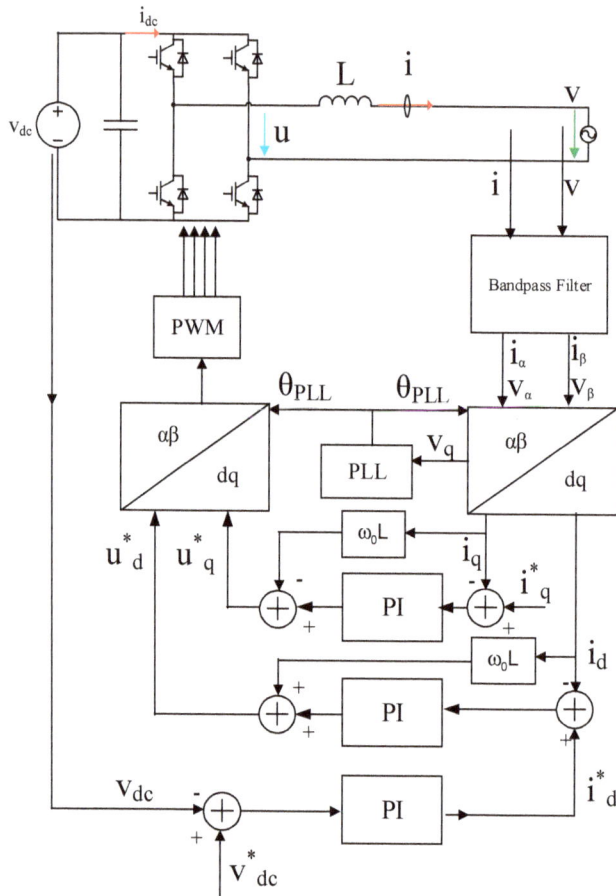

Figure 8. The block diagram of a typical synchronous rotating frame based current control inverter.

As shown in Figure 8, the output phase angle of the PLL θ_{PLL} affects the results of the dq transformation and inverse dq transformation. With a small-signal perturbation, due to the PLL output difference δ_{PLL}, there is an influence on the value associated with the dq transformation. In order to present this influence, a small signal model is shown in Figure 9.

In Figure 9, the hardware system (in the grey box) is the same as in Figure 6. $G_{iPL}L(s)$ is the transfer function matrix from the grid voltage disturbance to the inverter current disturbance in the controller, as expressed in Equation (19). $G_{uPLL}(s)$ is the transfer function matrix from the grid voltage disturbance to the inverter output reference voltage disturbance in the controller, as expressed in Equation (20). $G_{deco}(s)$ is the decoupling control matrix and $G_C(s)$ is the current regulator matrix, as expressed in Equations (21) and (22), separately [27].

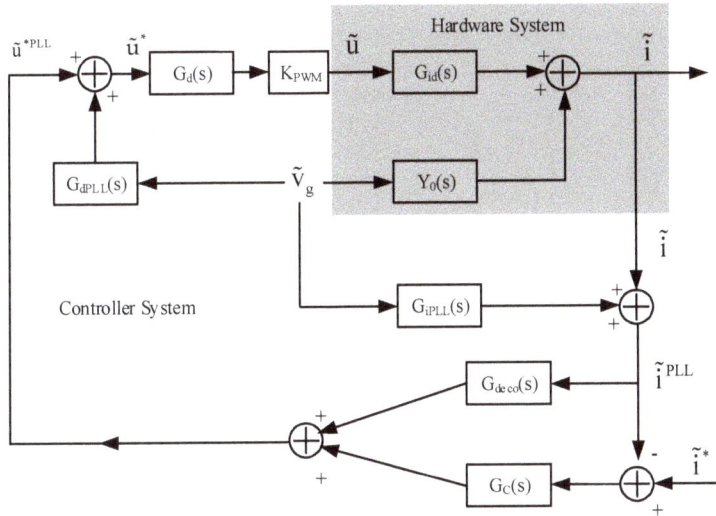

Figure 9. The block diagram of a typical synchronous rotating frame based current control inverter.

$$\tilde{i}_{dq}^{PLL} = G_{iPLL}(s)\tilde{v}_{dq} + \tilde{i}_{dq} \tag{19}$$

$$\tilde{u}_{dq}^{*PLL} = G_{uPLL}\tilde{v}_{dq} + \tilde{u}_{dq}^{*} \tag{20}$$

$$G_{deco}(s) = \begin{bmatrix} 0 & -\omega L \\ \omega L & 0 \end{bmatrix} \tag{21}$$

$$G_C(s) = \begin{bmatrix} K_{P-C} + \frac{K_{I-C}}{s} & 0 \\ 0 & K_{P-C} + \frac{K_{I-C}}{s} \end{bmatrix} \tag{22}$$

By applying the same derivation method in Section 3.1, Y_0 and G_{id} can be derived, as expressed in Equation (23).

$$\begin{cases} Y_0(s) = -\frac{1}{L(\omega^2+s^2)} \begin{bmatrix} s & \omega \\ -\omega & s \end{bmatrix} \\ G_{id}(s) = \frac{1}{L(\omega^2+s^2)} \begin{bmatrix} s & \omega \\ -\omega & s \end{bmatrix} \end{cases} \tag{23}$$

Small-signal perturbation is added to Equations (19), (20), and (10) to derive Equations (24) and (25).

$$\begin{bmatrix} I_d^{PLL} + \tilde{i}_d^{PLL} \\ I_q^{PLL} + \tilde{i}_q^{PLL} \end{bmatrix} \approx \begin{bmatrix} 1 & \tilde{\delta}_{PLL} \\ -\tilde{\delta}_{PLL} & 1 \end{bmatrix} \cdot \begin{bmatrix} I_d + \tilde{i}_d \\ I_q + \tilde{i}_q \end{bmatrix} \tag{24}$$

$$\begin{bmatrix} U_d^{*PLL} + \tilde{u}_d^{*PLL} \\ U_q^{*PLL} + \tilde{u}_q^{*PLL} \end{bmatrix} \approx \begin{bmatrix} 1 & \tilde{\delta}_{PLL} \\ -\tilde{\delta}_{PLL} & 1 \end{bmatrix} \cdot \begin{bmatrix} U_d^* + \tilde{u}_d^* \\ U_q^* + \tilde{u}_q^* \end{bmatrix} \tag{25}$$

By eliminating the steady state values in Equations (24) and (25), we obtain Equations (26) and (27).

$$
\begin{bmatrix} \tilde{i}_d^{PLL} \\ \tilde{i}_q^{PLL} \end{bmatrix} \approx \begin{bmatrix} \tilde{i}_d + \tilde{\delta}_{PLL} I_q \\ \tilde{i}_q - \tilde{\delta}_{PLL} I_d \end{bmatrix}
\tag{26}
$$

$$
\begin{bmatrix} \tilde{u}_d^{*PLL} \\ \tilde{u}_q^{*PLL} \end{bmatrix} \approx \begin{bmatrix} \tilde{u}_d^* + \tilde{\delta}_{PLL} U_q^* \\ \tilde{u}_q^* - \tilde{\delta}_{PLL} U_d^* \end{bmatrix}
\tag{27}
$$

By putting Equation (15) into Equations (26) and (27), Equations (28) and (29) are obtained.

$$
\begin{bmatrix} \tilde{i}_d^{PLL} \\ \tilde{i}_q^{PLL} \end{bmatrix} \approx \begin{bmatrix} 0 & G_{PLL} I_q \\ 0 & -G_{PLL} I_d \end{bmatrix} \cdot \begin{bmatrix} \tilde{v}_d \\ \tilde{v}_q \end{bmatrix} + \begin{bmatrix} \tilde{i}_d \\ \tilde{i}_q \end{bmatrix}
\tag{28}
$$

$$
\begin{bmatrix} \tilde{u}_d^* \\ \tilde{u}_q^* \end{bmatrix} \approx \begin{bmatrix} 0 & -G_{PLL} U_q^* \\ 0 & G_{PLL} U_d^* \end{bmatrix} \cdot \begin{bmatrix} \tilde{v}_d \\ \tilde{v}_q \end{bmatrix} + \begin{bmatrix} \tilde{u}_d^{*PLL} \\ \tilde{u}_q^{*PLL} \end{bmatrix}
\tag{29}
$$

According to Figure 9 (the relationship from the grid voltage to the inverter output current) by considering the influence of PLL, the inverter output admittance Y_{inv} and the current loop control transfer function H_C can be derived, as shown in Equations (30)–(32), respectively.

$$
i = H_C(s)i^* + Y_{inv}(s)v
\tag{30}
$$

$$
Y_{inv} = \frac{Y_0(G_d K_{PWM}(G_{iPLL}(G_{deco} - G_c) + G_{dPLL}) - I)}{I + G_d Y_0 K_{PWM}(G_{deco} - G_c)}
\tag{31}
$$

$$
H_C = \frac{Y_0 G_C G_d K_{PWM}}{I + Y_0 G_d K_{PWM}(G_{deco} - G_C)}
\tag{32}
$$

I represents the identity matrix.

4. Simulation Verification

The system as described above is analyzed analytically in the time domain (EMT-simulation) as well as in the frequency domain, using the equations from Section 3. The parameters in Table 1 are applied during the output admittance calculation and simulation. In this article, the influence of the d-axis component on the d-axis component of the inverter is analyzed. The analysis methods of other influences are similar to this and will not be repeated. Parameters were selected to achieve the same behavior as measured.

Table 1. The parameters of the inverters.

Parameter	DQ	PR
Rated power		3.3 kW
Rated voltage		230 V
Filter inductor		3.5 mH
PLL bandwidth		25 Hz
Sampling frequency		10 kHz
KP-C	2.46	/
KI-C	548.81	/
KP-PR	/	34.99
KR-PR	/	400
ωc	/	π

Verification

We put the parameters in Table 1 into Equations (18) and (31) to illustrate the bode diagram of the inverter output impedance. In the simulation, a small harmonic voltage of 100–5000 Hz is added to the grid voltage, the inverter output current is measured, and the output impedance of the inverter is obtained after the calculation of the measured data, as shown in Figure 10.

As shown in Figure 10, the analytic calculation results (Equations (18) and (31)) and the results from the simulation match very well. The accuracy of the linearized analytical model in the frequency domain has been verified.

The real part of the inverter output admittance $\text{Re}\{Y(j\omega)\}$ is illustrated in Figure 11.

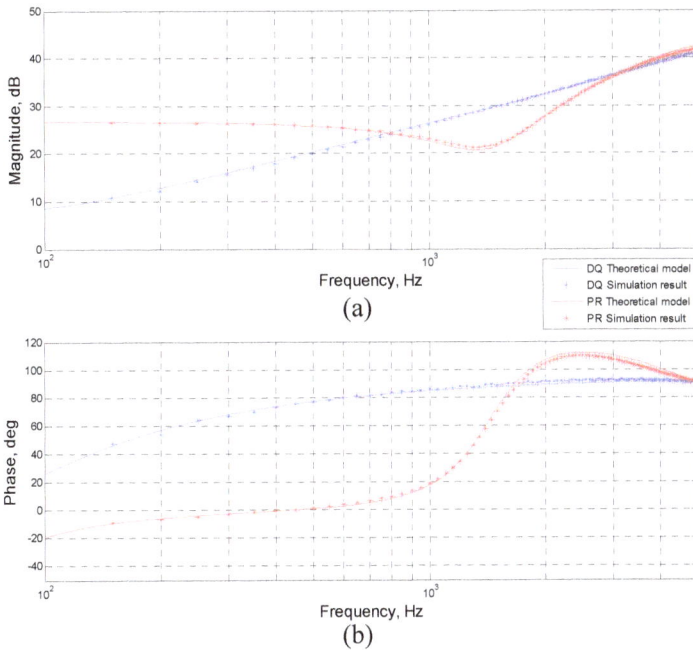

Figure 10. The Bode diagram of the inverter output impedance. (**a**) Impedance amplitude (**b**) Impedance phase angle; the solid line for theoretical calculation; * for simulation result; red for dq-based control and blue for PR-based control.

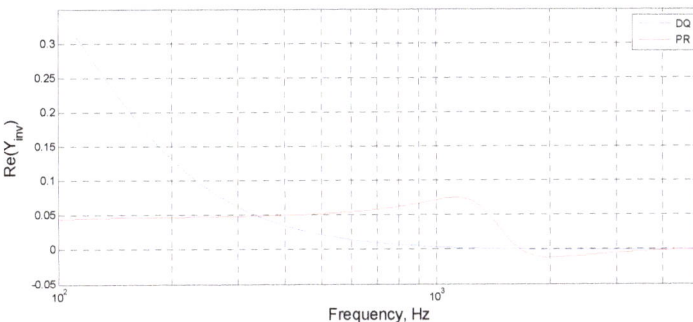

Figure 11. The real part of the inverter output admittance; blue for the dq-based control method and red for the PR-based control method.

135

As shown in Figure 11, both the dq-based control method (blue line) and PR-based control method (red line) enter the range with the negative real part of the output admittance near about 1700 Hz. This means that both control methods will be not stable if the resonance point of the grid is above 1700 Hz (Critical resonant frequency).

Verification will be performed in the simulation. In the simulation, the grid impedance consists of a CL circuit. The filter inductance of the inverter L, the grid capacitance Cg and the grid inductance Lg together form an LCL resonant circuit. Lg is given as 49.5 μH, corresponding to a 400 m cable length. The total grid capacitance, taking into account all the LV-cables in the analyzed system, will be estimated so that a specified resonance frequency is achieved (Equation (33)).

$$C_g = \frac{L + L_g}{4LL_g(\pi f_r)^2} \tag{33}$$

The resonant frequencies are set to 1530 Hz and 1730 Hz, respectively. To demonstrate the instability effect, the time domain simulation is started without capacitance and after 0.3 s, the capacitance is connected, forming an LCL circuit. The simulation results are shown in Figure 12.

As shown in Figure 12a,c, the inverter-grid system with the dq-based control method oscillates slightly after 0.3 s since the Re{$Y_{inv}(j\omega)$} of the inverter with the dq-based control method is near 0 in a large frequency range (1000 Hz~5000 Hz), as shown in Figure 11.

As shown in Figure 12b, when f_r = 1530 Hz is less than the critical resonant frequency (1700 Hz), the inverter-grid system with the PR-based control method is stable after 0.3 s. As shown in Figure 12f, when f_r = 1730 is larger than the critical resonant frequency (1700 Hz), the system is unstable after 0.3 s.

By comparing Figure 12c,d with the same system resonant frequency, the dq-based control method is critically stable and the PR-based control method is resonance instable. Since the dq-based control method (Figure 11a, blue line) is always above the PR-based control method (Figure 11a, red line) after entering the range with the negative real part of the output admittance, this means that the dq-based control method has a larger stability margin than the PR-based control method.

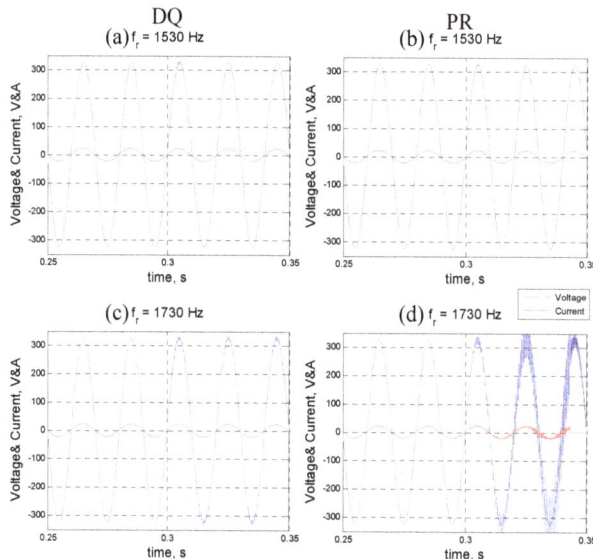

Figure 12. The simulation waveform result of (**a**) dq-based control method with a 1530 Hz resonant frequency; (**b**) PR-based control method with a 1530 Hz resonant frequency; (**c**) dq-based control method with a 1730 Hz resonant frequency; (**d**) PR-based control method with a 1730 Hz resonant frequency.

5. Analysis of the Influence of the Control Parameter and Improvement Methods

5.1. Analysis of the Influence of the Control Parameter

Figure 13 illustrates the influence of the PLL bandwidth on Re{Yinv(jw)}, the abscissa shows the frequency, the ordinate shows the PLL bandwidth and different colors represent Re{Yinv(jw)}. The parameters in Table 1 are used to generate Figure 13.

In Figure 13, the abscissa represents the frequency of Re{Y_{inv}(jw)}, the ordinate represents the PLL bandwidth and the color in the figure represents the amplitude of the Re{Y_{inv}(jw)}, with warm colors representing larger amplitudes and cool colors representing smaller amplitudes. Dark blue represents the negative amplitude of the Re{Y_{inv}(jw)}. The blue vertical dotted line in Figure 13a and red vertical dotted line in Figure 13b represent the 1700 Hz line.

As shown in Figure 13a, the increase of the PLL bandwidth has no effect on the Re{Y_{inv}(jw)} of the dq-based control method. With the increasing PLL bandwidth, the negative area of the Re{Y_{inv}(jw)} of the PR-based control method expands at a low frequency (Figure 13b). The positive area shrinks even though the amplitude increases. In general, the stability frequency area of the system becomes smaller.

In order to specifically analyze the influence of the control parameters on a certain frequency, 1700 Hz is selected as the target frequency for analysis in this article. The increased PLL bandwidth (>2000 Hz) will cause the Re{Y_{inv}(jw)} of the PR-based control method to enter a stable area at 1730 Hz (Figure 13b). Thus, when the resonant frequency of the grid impedance is around 1700 Hz, the inverter-grid system can still be stable.

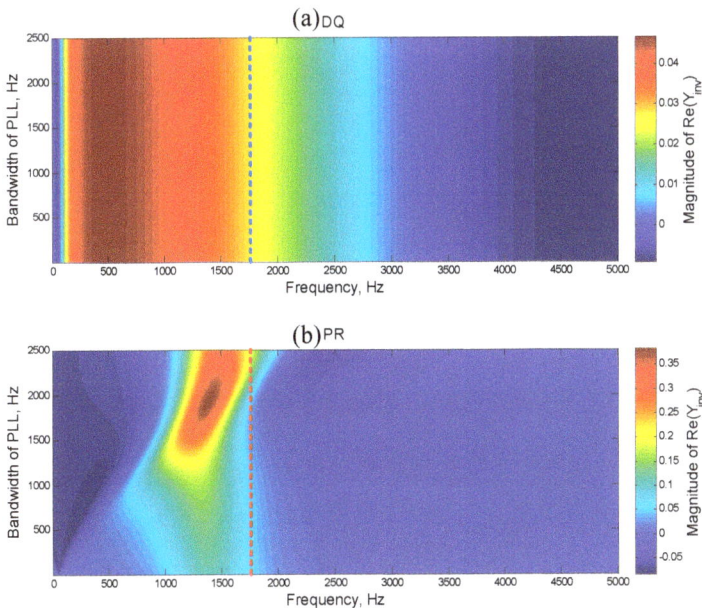

Figure 13. The influence from the PLL bandwidth to Re{Yinv(jw)}, full band. (**a**) dq-based control method; (**b**) PR-based control method.

In Figure 14, the abscissa represents the frequency of Re{Y_{inv}(jw)}, the ordinate represents the PI bandwidth for the current control loop of the dq-based control method (Figure 14a), and the proportion coefficient KP-PR of the PR-based control method (Figure 14b). The red vertical dotted line in Figure 14a and the blue vertical dotted line in Figure 14b represents the 1700 Hz line.

By increasing the PI bandwidth for the current control loop of the dq-based control method, the negative area of the Re{Y_{inv}(jω)} expands at a low frequency, the positive area moves to a higher frequency area, and the amplitude peak of Re{Y_{inv}(jω)} also decreases, as shown in Figure 14a. For the PR-based control method, by increasing the proportion coefficient K_{P-PR}, the positive area (Figure 14b, in the red dotted box) shrinks. Thus, increasing the PI control bandwidth or proportion coefficient will result in a reduction in the system's stable frequency area.

By increasing the PI bandwidth for the current control loop of the dq-based control method, the negative area of the Re{Y_{inv}(jω)} expands at a low frequency, the positive area moves to a higher frequency area, and the amplitude peak of Re{Y_{inv}(jω)} also decreases, as shown in Figure 14a. For the PR-based control method, by increasing the proportion coefficient K_{P-PR}, the positive area (Figure 14b, in the red dotted box) shrinks. Thus, increasing the PI control bandwidth or the proportion coefficient will result in a reduction in the system's stable frequency area.

When the frequency is 1700 Hz, the increase of the PI bandwidth for the current control loop of the dq-based control method cannot significantly change the amplitude of the Re{Y_{inv}(jω)}, as shown in Figure 14a (the red line). The increase of the proportion coefficient K_{P-PR} for the current control loop of the PR-based control method will let Re{Y_{inv}(jω)} < 0 if K_{P-PR} > 34. In other words, when K_{P-PR} < 34, if the resonant frequency of the grid impedance is near 1700 Hz, the inverter will be stable.

In addition to reducing the PLL bandwidth and current loop control bandwidth, increasing the filter inductance of the inverter or reducing the delay of the controller can also extend the positive range of Re{Y_{inv}(jω)}. However, these parameters are limited by the inverter's hardware and controller performance, it is not convenient for them to be changed in practical operations.

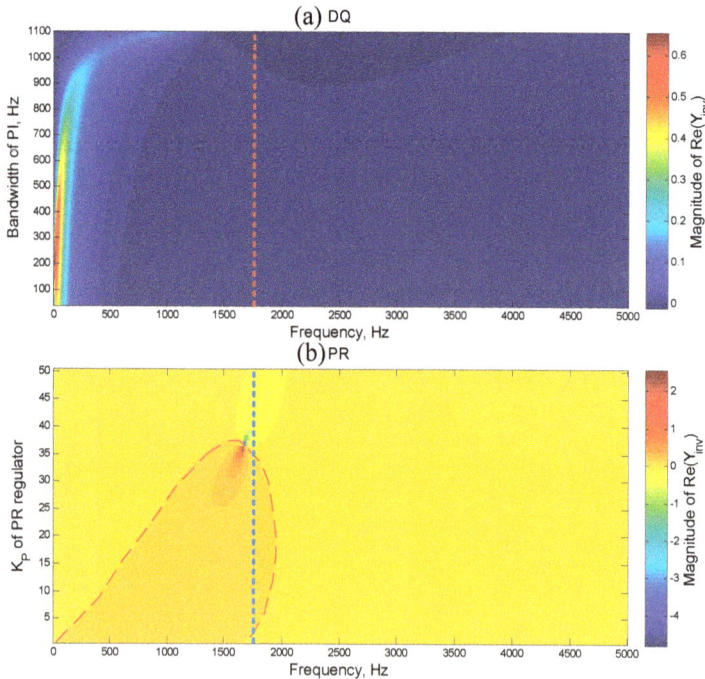

Figure 14. The influence from the current control parameters to Re{Yinv(jω)}, full band. (**a**) dq-based control method; (**b**) PR-based control method.

5.2. Improvement Methods

As discussed in Section 5.1, choosing a lower PLL bandwidth and current control bandwidth let $\text{Re}\{Y_{inv}(j\omega)\}$ expand its positive area to increase the stability margin of the inverter. In the case that the resonant frequency of the grid impedance is unknown, the method of reducing the control bandwidth can maximize the robustness of the inverter.

Another way to improve the stability is to install the filter on the output side of the inverter to eliminate the influence of the grid impedance resonance on the inverter. If the resonant frequency of the grid impedance is known, a low-pass filter with a cutoff frequency below the resonant frequency or a notch filter can be connected on the output side of the inverter.

To demonstrate the effect of these two improved methods, a time domain simulation with the same process and parameters of Figure 12 is carried out. The resonant frequency of the grid impedance is set to 1730 Hz, the same as in Figure 12c,d. The time domain simulation is started without the capacitance and after 0.3 s, the capacitance is connected, forming an LCL circuit. The simulation results are shown in Figure 15.

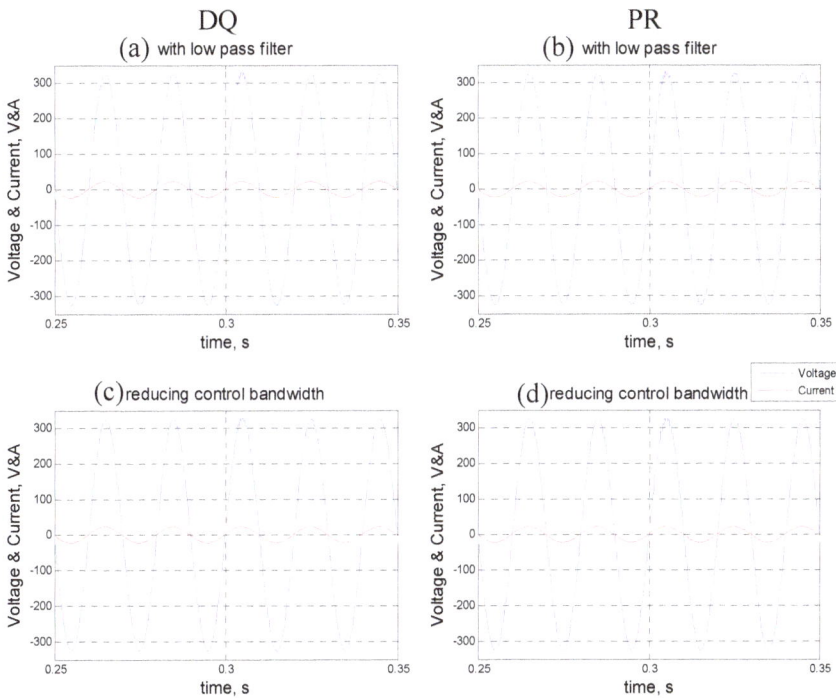

Figure 15. The simulation waveform result of (**a**) the dq-based control method with a low pass filter; (**b**) the PR-based control method with a low pass filter; (**c**) the dq-based control method with a reduced control bandwidth; (**d**) the PR-based control method with a reduced control bandwidth.

Since the resonant frequency of the grid impedance is 1730 Hz, a low-pass filter with a cutoff frequency of 850 Hz is connected to the output side of the inverter in the simulation. As shown in Figure 15a,b, the inverter-grid system is stable after 0.3 s. Comparing Figures 13c,d and 15a,b, the stability of the inverter-grid system with two different control methods can be seen to have significantly improved.

The simulation results for reducing the control bandwidth are shown in Figure 15c,d. The PLL bandwidth and current control bandwidth are simultaneously reduced to 20% of the original value (Table 1). As shown in Figure 13c,d, the inverter-grid system is stable after 0.3 s. Comparing Figures 12c,d and 15c,d, the stability of the inverter-grid system with the two different control methods have been significantly improved.

6. Conclusions

In this article, the resonance instability phenomenon that was found during tests of a photovoltaic charging station is analyzed. The interaction of a resonance situation in the grid impedance and non-passivity of the inverter in that frequency range causes the system formed by the inverter and grid to be non-stable.

The inverter output admittance determines whether the inverter is passive or not. In order to analyze the inverter output admittance, the single-phase inverter system is modelled, taking into account the two common control methods. These are, respectively, the PR-based current control and the synchronous rotating frame based current control methods (dq-based). In order to consider the effect of the PLL on the output admittance, a linear small signal model is used for the PLL modeling.

This article verified the modeling by simulation. The analytical model shows that the real part of the inverter output admittance is negative for frequencies above 1700 Hz. In case the resonant frequency of the grid impedance exceeds 1700 Hz, the inverter-grid system will be unstable. The results are confirmed by the time domain simulation.

This article further analyzes the influence of different control parameters on the inverter output admittance. The analysis shows that the PLL bandwidth and the control parameters of the current loop will have an influence on the output admittance of the inverter. By adjusting these parameters, the inverter-grid system may be prevented from being unstable at certain resonant frequencies.

On the basis of the above analysis, this article gives two methods to improve stability: connecting a low pass filter or reducing the control bandwidth. The simulation results show that both of these methods can significantly improve the stability of the inverter-power grid system.

Author Contributions: Writing–original draft preparation, formal analysis and software, Z.Z.; investigation and validation, Z.Z. and C.G.; conceptualization, resources, L.F., H.R and A.R; review and editing C.G., H.R., L.F. and A.R.; supervision, project administration and funding acquisition, A.R. and L.F.

Funding: This research has received funding from the European Union's Horizon 2020 research and innovation programme under the ERA-Net Smart Grids plus, grant number 646039, from the Netherlands Organisation for Scientific Research (NWO) and from BMVIT/BMWFW under the Energy der Zukunft programme.

Acknowledgments: Our project has received funding in the framework of the joint programming initiative ERA-Net Smart Grids Plus, with support from the European Union's Horizon 2020 research and innovation programme. We would like to thank University of Twente for supporting this research study, as part of the Solar bike and Smart Living Campus projects.

Conflicts of Interest: The authors declare no conflict of interest.

Disclaimer: The content and views expressed in this material are those of the authors and do not necessarily reflect the views or opinion of the ERA-Net SG+ initiative. Any reference given does not necessarily imply the endorsement by ERA-Net SG+.

References

1. Biresselioglu, M.E.; Demirbag Kaplan, M.; Yilmaz, B.K. Electric mobility in Europe: A comprehensive review of motivators and barriers in decision making processes. *Transp. Res. Part A Policy Pract.* **2018**, *109*, 1–13. [CrossRef]
2. Fathabadi, H. Novel grid-connected solar/wind powered electric vehicle charging station with vehicle-to-grid technology. *Energy* **2017**, *132*, 1–11. [CrossRef]
3. Nunes, P.; Figueiredo, R.; Brito, M.C. The use of parking lots to solar-charge electric vehicles. *Renew. Sustain. Energy Rev.* **2016**, *66*, 679–693. [CrossRef]

4. Redpath, D.A.G.; McIlveen-Wright, D.; Kattakayam, T.; Hewitt, N.J.; Karlowski, J.; Bardi, U. Battery powered electric vehicles charged via solar photovoltaic arrays developed for light agricultural duties in remote hilly areas in the Southern Mediterranean region. *J. Clean. Prod.* **2011**, *19*, 2034–2048. [CrossRef]

5. Ji, S.; Cherry, C.R.; Han, L.D.; Jordan, D.A. Electric bike sharing: Simulation of user demand and system availability. *J. Clean. Prod.* **2014**, *85*, 250–257. [CrossRef]

6. Apostolou, G.; Reinders, A.; Geurs, K. An Overview of Existing Experiences with Solar-Powered E-Bikes. *Energies* **2018**, *11*, 2129. [CrossRef]

7. Apostolou, G.; Guers, K.; Reinders, A. Technical performance and user aspects of solar powered e-bikes—Results of a field study in The Netherlands. Presented at the DIT–ESEIA Conference on Smart Energy Systems in Cities and Regions, Dublin, Ireland, 10–12 April 2018.

8. Zhao, M.; Yuan, X.; Hu, J.; Yan, Y. Voltage Dynamics of Current Control Time-Scale in a VSC-Connected Weak Grid. *IEEE Trans. Power Syst.* **2016**, *31*, 2925–2937. [CrossRef]

9. Krein, P.T.; Bentsman, J.; Bass, R.M.; Lesieutre, B.L. On the use of averaging for the analysis of power electronic systems. *IEEE Trans. Power Electron.* **1990**, *5*, 182–190. [CrossRef]

10. Sun, J. AC power electronic systems: Stability and power quality. In Proceedings of the 2008 11th Workshop on Control and Modeling for Power Electronics, Zurich, Switzerland, 17–20 August 2008; pp. 1–10.

11. Kroutikova, N.; Hernandez-Aramburo, C.A.; Green, T.C. State-space model of grid-connected inverters under current control mode. *IET Electr. Power Appl.* **2007**, *1*, 329. [CrossRef]

12. Bengtsson, T.; Bickel, P.; Li, B. Curse-of-dimensionality revisited: Collapse of the particle filter in very large scale systems. In *Institute of Mathematical Statistics Collections*; Institute of Mathematical Statistics: Beachwood, OH, USA, 2008; pp. 316–334. ISBN 978-0-940600-74-4.

13. Sun, J. Impedance-Based Stability Criterion for Grid-Connected Inverters. *IEEE Trans. Power Electron.* **2011**, *26*, 3075–3078. [CrossRef]

14. Willems, J.C. Dissipative Dynamical Systems. *Eur. J. Control* **2007**, *13*, 134–151. [CrossRef]

15. Hu, H.; Tao, H.; Blaabjerg, F.; Wang, X.; He, Z.; Gao, S. Train–Network Interactions and Stability Evaluation in High-Speed Railways–Part I: Phenomena and Modeling. *IEEE Trans. Power Electron.* **2018**, *33*, 4627–4642. [CrossRef]

16. Pan, P.; Hu, H.; Yang, X.; Blaabjerg, F.; Wang, X.; He, Z. Impedance Measurement of Traction Network and Electric Train for Stability Analysis in High-Speed Railways. *IEEE Trans. Power Electron.* **2018**, *33*, 10086–10100. [CrossRef]

17. Reinders, A.; de Respinis, M.; van Loon, J.; Stekelenburg, A.; Bliek, F.; Schram, W.; van Sark, W.; Esteri, T.; Uebermasser, S.; Lehfuss, F.; et al. Co-evolution of smart energy products and services: A novel approach towards smart grids. In Proceedings of the 2016 Asian Conference on Energy, Power and Transportation Electrification, Singapore, 25–27 October 2016; pp. 1–6.

18. Canadian Solar. *All-Black CS6K-270-275-280 M*. Available online: https://www.canadiansolar.com/downloads/datasheets/v5.531/canadian_solar-datasheet-allblack-CS6K-M-v5.531en.pdf (accessed on 30 November 2018).

19. olax Power. *X1 Series User Manual*. Available online: https://www.solaxpower.com/wp-content/uploads/2017/01/X1-Mini-Install-Manual.pdf (accessed on 30 November 2018).

20. Ciobotaru, M.; Agelidis, V.; Teodorescu, R. Line impedance estimation using model based identification technique. In Proceedings of the 2011 14th European Conference on Power Electronics and Applications, Birmingham, UK, 30 August–1 September 2011; pp. 1–9.

21. Guo, X.; Wu, W.; Chen, Z. Multiple-Complex Coefficient-Filter-Based Phase-Locked Loop and Synchronization Technique for Three-Phase Grid-Interfaced Converters in Distributed Utility Networks. *IEEE Trans. Ind. Electron.* **2011**, *58*, 1194–1204. [CrossRef]

22. Pendharkar, I. A generalized Input Admittance Criterion for resonance stability in electrical railway networks. In Proceedings of the 2014 European Control Conference (ECC), Strasbourg, France, 24–27 June 2014; pp. 690–695.

23. Ban, M.; Shen, K.; Wang, J.; Ji, Y. A novel circulating current suppressor for modular multilevel converters based on quasi-proportional-resonant control. *China Acad. J. Electron. Publ. House* **2014**, 85–89. [CrossRef]

24. Lu, M.; Wang, X.; Loh, P.C.; Blaabjerg, F.; Dragicevic, T. Graphical Evaluation of Time-Delay Compensation Techniques for Digitally Controlled Converters. *IEEE Trans. Power Electron.* **2018**, *33*, 2601–2614. [CrossRef]

25. Mattavelli, P.; Polo, F.; Dal Lago, F.; Saggini, S. Analysis of Control-Delay Reduction for the Improvement of UPS Voltage-Loop Bandwidth. *IEEE Trans. Ind. Electron.* **2008**, *55*, 2903–2911. [CrossRef]
26. Wu, H.; Ruan, X.; Yang, D. Research on the stability caused by phase-locked loop for LCL-type grid-connected inverter in weak grid condition. *Zhongguo Dianji Gongcheng Xuebao/Proc. Chin. Soc. Electr. Eng.* **2014**, *34*. [CrossRef]
27. Wen, B.; Boroyevich, D.; Burgos, R.; Mattavelli, P.; Shen, Z. Analysis of D-Q Small-Signal Impedance of Grid-Tied Inverters. *IEEE Trans. Power Electron.* **2016**, *31*, 675–687. [CrossRef]

MDPI

St. Alban-Anlage 66

4052 Basel

Switzerland

Tel. +41 61 683 77 34

Fax +41 61 302 89 18

www.mdpi.com

Applied Sciences Editorial Office

E-mail: applsci@mdpi.com

www.mdpi.com/journal/applsci

www.ingramcontent.com/pod-product-compliance
Lightning Source LLC
Chambersburg PA
CBHW051906210326
41597CB00033B/6047